THE POWER OF MUSIC

THE POWER OF MUSIC
Psychoanalytic Explorations

Roger Kennedy

PHOENIX
PUBLISHING HOUSE
firing the mind

First published in 2020 by
Phoenix Publishing House Ltd
62 Bucknell Road
Bicester
Oxfordshire OX26 2DS

British Library Cataloguing in Publication Data

A C.I.P. for this book is available from the British Library

ISBN-13: 978-1-912691-73-9

Typeset by vPrompt eServices Pvt Ltd, India

Printed in the United Kingdom

www.firingthemind.com

Contents

About the author

Dr Roger Kennedy is a consultant child and adolescent psychiatrist and an adult psychoanalyst. He was an NHS consultant in charge of the Family Unit at the Cassel Hospital for nearly thirty years, before going totally into private practice ten years ago. He was chair of the Child and Family Practice in Bloomsbury and is still a director there. His work includes being a training analyst and seeing adults for analysis and therapy, as well as children, families, and parents at his clinic. He is a past president of the British Psychoanalytical Society, and is well known as an expert witness in the family courts. He has had thirteen previous books published on psychoanalysis, interdisciplinary studies, and child, family, and court work, as well as many papers.

CHAPTER ONE

Overture

This book explores, with the help of a psychoanalytic perspective, some of the ways one can understand how music affects us, as we listen to it, and what it is about music that affects us so profoundly and powerfully, often at a deep emotional as well as at an intellectual level. There tend to be two ways of looking at the relationship between music and emotion, the "cognitivist" view which considers that music merely expresses emotions without inducing them, and the "emotivist" view which considers that music induces emotions in listeners. My own take on this is that there is considerable and clear evidence from many sources that music induces brief and sustained emotions of various kinds, and I also take the view that music has a considerable power to express a whole range of emotions, from the most basic and short-lasting emotions to complex and more long-lasting emotions, including those usually involved with spiritual or "other worldly" experiences. Emotion is an integral aspect of musical experience; music has the power to take us on an emotional and intellectual journey, transforming the listener along the way; the aim of the book is to examine the nature of this journey.

Evidence from neuroscience indicates that music acts on a number of different brain sites, and that the brain is likely to be hard-wired for musical perception and appreciation; this offers some kind of neurological substrate for musical experiences, or a parallel mode of explanation for music's multiple effects on individuals and groups.

After various excursions into early mother–baby experiences, evolutionary speculations, and neuroscientific findings, my main emphasis will be that it is the *intensity* of the artistic vision which is responsible for music's power; that intense vision invites the viewer or the listener into the orbit of the work, engaging us to respond to the particular vision in an essentially intersubjective relationship between the work and the observer or listener. This is the area of what we might call the human soul. Music can be described as having soul when it hits the emotional core of the listener. And, of course, there is "soul music", whose basic rhythms reach deep into the body to create a powerful feeling of aliveness. One can truly say that music, of all the arts, is the most able to give shape to the elusive human subject or soul.

In addition, so much of musical enjoyment comes from surrendering to the music, not just because the sounds themselves elicit pleasure, not just because of the effect of the specific emotional content and the play of musical tensions, but because we can lose ourselves in the music while also remaining ourselves; these are the moments of pleasurable illusion. As I shall explore, music has the power in performance to touch the intermediate area between subject and other, giving us the illusion of wholeness, however briefly, where subject and other merge.

This opening chapter is, as befits its title, an overview of the kind of themes I will develop in more detail as the book unfolds. I will cover a number of different ways of understanding music, including psychological, social, cultural, and biological and neuroscientific approaches, as music perception and understanding and musical studies cover so many aspects of human life; no single discipline can do justice to music's complexity if one is to have a sense of the whole musical experience, even if one has to break up the whole experience into various elements for the purposes of clarification.[1]

The issues raised here will have some relationship to psychoanalytic understanding and listening, as after all psychoanalysis is a listening discipline; its bedrock is listening to the patient's communications. While of course there are significant differences between understanding of, and listening to, a musical performance and a patient in a consulting room, there is also, as I shall explore, some common ground.

In psychoanalytical listening, one is listening simultaneously to the "surface" and the "depth" of the patient's communications, to both the conscious and underlying unconscious stream of thoughts and feelings.

There are some loose parallels with this kind of listening and musical analysis, particularly the kind that looks beneath the surface of the musical "foreground" to the underlying deep structures of the "background".[2] However, such analysis is a highly sophisticated and intellectual exercise, while listening psychoanalytically and musically are also intensely emotional experiences.

Analytic listening, however intellectually taxing at times, also entails a responsive, receptive, or affective kind of listening, more like trying to make sense of the shape of the communications. During a session the analyst may become immersed in the flow of the patient's material. "As in listening to music, one may follow the melody line, the obbligato, the counterpoint. The analyst is free to move from one line to the other, to hear them all simultaneously."[3]

This kind of listening has also been described as a kind of musical "reverie" by Riccardo Lombardi,[4] which can arise in the analyst particularly during intense emotional exchanges. Lombardi suggests that musical associations have a peculiar capacity of bridging the gap between the concrete and the abstract, between body and mind, the non-symbolic and the symbolic, the internal and the external. He shows how the use of the analyst's countertransferential musical reverie can be used for clinical purposes. As he describes, music is connected with both the concrete world of bodily sensations and the symbolic expression of culture, and may then be an important transitional phenomenon in communication between patient and analyst on both conscious and unconscious levels. He gives two clinical examples where each of the patients' awareness of the passage of time, associated with the analyst's internal musical associations and reverie, made it possible for them to give them access to more fluid and less rigid emotions, with a reduction in their symptoms and with a loosening up of their rigid defences.

Theodor Reik, who saw music as intimately linked to emotions and psychic reality, had already pointed out, with many clinical examples, how musical associations arising in the analyst's mind can be of great help in the understanding of the patient's communications. "The tunes occurring to the analyst during sessions with patients are preconscious messages of thoughts that are not only meaningful, but also important for the understanding of the emotional situation of the patient … The tunes stand in the service of the agents responsible for the communication between the unconscious of two persons."[5]

To present just one example given by Reik:[6] a patient has a dream. *She is in the bath and is worried because she has forgotten to take off her watch which could be ruined if it gets wet.* There were no helpful associations to the dream. In the pause between her report of the dream and the following sentences she spoke, a long-forgotten tune came to Reik's mind, which he then realised he had not heard since childhood. The title was *The Watch* (by Karl Löwe). He recalled later the first lines: "Wherever I go, I carry a watch with me always, and only need look whenever I'd know the time of day." The watch meant the human heart. Reik then recalled the phrase that Viennese girls used to say, referring to their periods, that "With me it is punctual as a watch." At the next session, the patient referred to her dream and that she had forgotten to put in her diaphragm after her bath, and was worried that intercourse might have led to a pregnancy.

Francis Grier, a composer and a psychoanalyst, in his paper on musicality and the consulting room,[7] describes in detail how attention to the specifically musical dimensions of non-verbal relating in the analytic encounter reaches deep and emotionally resonant areas of the patient's psyche, which verbal language cannot easily reach. For example, attention to the quality, the resonances of the patient's voice, and how this may evoke reactions in the analyst, that is, attention to the "duet" between analyst and patient, may lead to useful analytic work. He gives a detailed example of how he was affected by the grating quality of his patient's voice, and yearned to experience other pitches and resonances, instead of being trapped in the session with her highly dissonant high-pitched communications. Analytic work on his countertransference feelings about the lack of harmony between the content of her speech and its expression led to deep work on her early dissonant experiences, an idealised but dysfunctional relationship with her mother, with an early over-reliance on her father as the only reliable figure.

Although Grier never spoke to his patient in musical terms, he describes that when he allowed himself privately to free associate musically about the quality of her voice and its dissonance, it freed him up interpretatively to get through what had become an impasse in the analytic work, so that in the course of time, in a way he did not quite understand logically, the situation between them moved. This corresponded to a change in his patient's voice, noticed by her friends and family; its pitch dropped, its expressiveness broadened, its unnaturally loud and forced dynamic began to quieten, and she was able to communicate emotion through more dynamic and harmonic means.

In the next chapter I will take up how the quality of the mother's voice impacts on the baby's development, providing an essential scaffolding for emotionality, and how a distorted or deadened voice can create emotional disturbances with far-reaching consequences.

Julie Nagel, in her book *Melodies of the Mind*,[8] also examines how the non-verbal qualities of music itself, its rhythm, pitch, tonality, form, and dynamics, can lead directly to understanding the patient's emotional life at a deep level. As she describes, music is a powerful point of entry into affect, meaning, preverbal representations and unconscious processes. In the book, she mainly examines the relationship between music and psychoanalysis by analysing various pieces of music. In other work,[9] she has written about how to help musicians with stage fright, focusing for example on shame dynamics, and fears about psychic and bodily disintegration. For a performer, particularly a soloist or an actor, neither of whom can escape the gaze of the audience, shame, linked to fears about being *seen*, can have a paralysing effect on their ability to function.

One could say that every patient has their own music, but that every analyst and patient encounter creates a music of its own. The analyst is thus engaged with listening to both the patient and to themselves and to their own responses to what the patient brings. One could say there are different kinds of music going on in the analytic situation. The patient may attempt to communicate an inarticulate fantasia, or maybe a suspiciously articulate composition. Then there is music that goes on in the analyst as he or she listens, as Reik, Lombardi, and Grier show, and in the way that interpretations are composed. And then there is the music made jointly in the analysis, a form of duet.

Michael Parsons[10] makes the point that the analyst's listening involves listening in two dimensions at once—externally to the patient and internally to what is stirred up by listening to the patient. That internal listening involves a certain kind of receptivity to the unconscious, which seems to have parallels with listening to music. It is, as Parsons also points out,[11] that kind of receptiveness which comes in when listening to poetry. With the latter, one may be making out the meaning of their words, but allowing the words to have an emotional impact is also an important part of the experience and one which may be difficult to tie down. This is similar to what Reik described as "hearing oneself", by which he meant the analyst using the unconscious as a receiving apparatus.[12] Parsons describes how being receptive to the "internal" music aroused in a listening analyst helps the analyst understand the external music that is the patient.

In trying to explore this listening process, it may help to consider the complexities of musical listening where of course dissonance and consonance are in constant interaction. Listening to music, being affected by the music one hears, involves a number of different elements; there is a network of human subjects engaged in complex acts of communication and interpretation involving both the intellect and emotion.

However, just focusing on the element of arousal, it seems clear that music induces a general state of arousal, heightened awareness, and expectations. In an analytic session, the particular quality of arousal may well have intimate connections with the patient's earliest object relations, how, for example, the early vitality affects between mother and baby were managed, as I shall explore in the next chapter.

That music has a profound effect on us is not in doubt. It affects us at various levels, emotionally, intellectually, and bodily to give us considerable pleasure. Music has a dual aspect—it touches individuals, reaching their innermost being, while it is also capable of bringing people together in a collective endeavour. Such collective power can be seen in many social situations, including music's role in facilitating group identity in young people, or empowering groups to unite together.

Strictly speaking, one should make a distinction between music as "humanly organized sound",[13] a cultural and social construct, which often involves some kind of performance, whether that be in a concert hall, opera house, listening studio, or in a community setting, and "musicality", which is more about abilities that make music production, listening, and appreciation possible, or the mental processes that underlie musical behaviour and perception, which may or may not be innate; but of course, the performance of music does imply musicality.

Music can also convey areas of experience that language cannot reach. Tchaikovsky put this rather well in a letter to his patroness Nadezhda von Meck, soon after the rapid break-up of his disastrous marriage:

> I will tell you that more than once I tried lovingly to express in music the torment and, at the same time, the bliss of love. Whether I have succeeded I do not know—or, rather, I leave others to judge. I totally disagree with you when you say that music *cannot convey the all-embracing characteristics of the feeling of love*. I believe quite the

contrary—that *music alone* can do this. You say that here *words* are necessary. Oh no! It is precisely here that words are not necessary— and where they are ineffectual, the more eloquent language, that is music, appears in all its power.[14]

It is worth noting here not only Tchaikovsky's view of the *power of music* over words to convey human emotion, but that music has the unique ability, unlike language, to convey simultaneously contrary emotions—here he refers to the torment and bliss of love. In addition, Tchaikovsky in his book on harmony describes the aim of music as being "to picture the many various emotions of the soul".[15]

Thus, we are not just aroused by the effect of music, but music can convey to us complex and multiple meanings and focused and powerful and even mixed emotional states. It seems to involve simultaneously the deepest and earliest layers of the self as well as the most sophisticated.[16] Music, more than any other human activity, is "an exercise in timing. Perhaps it serves as a means of coordinating the temporal activities of widely distributed brain regions."[17] That is, music's ability to synchronise different mental functions underlies our deep need to create and participate in musical activities, as I shall discuss when looking at neuro-scientific studies of music.

Music can create a sense of wholeness or can confuse or even fragment us by its power; it can rouse us into battle or provide laments for the fallen dead; it can excite us with rhythms and harmonies, make us feel sexually aroused, or soothe our troubled souls with elegies, bring us into intimate connection with others, or allow us to retreat into a private world of the imagination. Or music may play, as I shall explore later, a vital role in the ritual of a community as ethnomusical studies reveal in a variety of cultures, where music can be linked to complex social and cultural practices, including trance states, not unlike states of mind induced by pop concerts, brought on by the conjunction of powerful emotions and imagination, enhancing social connectedness as well as subverting ordinary ways of being and understanding.[18] Music is described by Shakespeare in *Twelfth Night* as the food of love as well as its enhancer. "In an Indian myth, a raga has the power to produce water, finally drowning the singer. In Blackfoot culture, singing the right song validates an activity. The Orpheus myth as well as Tamino of *The Magic Flute* performs music to make humans and animals and spirits behave."[19]

Music can have the extraordinary power to be able to evoke, through atmosphere, particular times, places, and people, as one can see with the Simon Bolivar Orchestra's capacity to evoke Latin American or Mexican worlds; or how a Grieg quartet, written with passion during his marriage crisis; or a Smetana quartet, written when he was going deaf and incorporating the sounds of tinnitus, can convey something new about human suffering and the ability to survive.

Music can thus have powerful links with human agency, and of all the arts it has the most immediate emotional impact on an audience, with an ability to release us from social restraints,[20] so that its source has not without reason been linked to "deep unconscious urges".[21]

Schopenhauer considered that music was the direct manifestation of the most intimate and secret *will* of the human being. As he put it: "[M]usic expresses in an exceedingly universal language, in a homogenous material, that is, in mere tones, and with the greatest distinctiveness and truth, the inner being, the in-itself, of the world, which we think of under the concept of the will."[22]

Music is the voice for Schopenhauer of the metaphysical will, which is why it appears to speak to us from the most ultimate depths, deeper by far than those accessible to other arts, while remaining unamenable to language or to intellectual understanding.[23]

For Hegel,[24] music is one of the "romantic arts". Music, together with painting and poetry, is able to give form to the objectless inner life of the human subject; it is the art of the soul that speaks to itself; its object is feeling, which, like sound, lives and develops in time. Though it must also be said that for Hegel music tends to lead to regression and a loss of conscious reason, rather than being life-enhancing.

Empirical studies of the effect of music[25] show that it has a wide variety of actions in a wide variety of contexts, but a constant theme seems to be that it has considerable transformative power for the human subject, transfiguring experience as well as mediating between the natural and the supernatural in trance and religious ecstasy. Viktor Zuckerkandl[26] even defines music as the "other power along with language which defines man as a spiritual being".

Music can bring a "delight quite removed from mere sense-impression, a delight which is felt to be bound up with very active powers of grasp and comprehension, and to be stirring very deep depths of emotion, and yet, totally, and at times it may be tantalisingly, isolated from the facts and interests of life".[27] Then we may at least enjoy the sheer beauty of the music's melodies and harmonies, and not care about what it may or may not mean, and just let the music resonate with our senses.

The response of the listener to music is a complex process, involving emotional and cognitive aspects. Music will tend to evoke some feelings in the listener of a general kind, often pleasurable, sometimes linked to memories, or even past memories of previous performances. The intensity of the listening experience may vary according to where the music is heard; listening to a live performance, for example, is usually more evocative than listening to a recording.

The *place* where the music is heard may also have particular relevance. For example, John Eliot Gardiner[28] describes how the sense of place was important while he and the Monteverdi Orchestra were recording all of Bach's cantatas, in a kind of musical pilgrimage. The original context in Bach's time was for performance in church during the cycle of the seasons and religious festivals. By taking them out of that context into the concert hall, something essential disappeared. The concert hall allows for focused listening, but all the many connections to the Lutheran year and hence their essential meaning are diminished. Eliot Gardiner arranged for performances in a variety of churches, mainly in Germany, in order to approximate the original context of the music. Musical meaning may, then, be linked to time and place or context; it requires a particular home.

Harry Christophers, in the BBC series on the history of sacred music,[29] describes how he and members of his group of singers, The Sixteen, discovered that the music of Léonin, the twelfth-century poet and musician attached to Notre-Dame, had a startling resonance with gothic architecture. The music created a different and more intense experience when sung in the cathedral setting; the music had evidently been composed with that particular space in mind.

The ethnomusicologist Steven Feld[30] showed how among the Kaluli people of Papua New Guinea, the forms of their music intimately reflect their surroundings and their culture. The Kalulis are sophisticated hunters and naturalists, intimately connected to their rainforest home, about which they have a deep knowledge, and which is reflected in their complex music. Much of their culture is organised around their knowledge and appreciation of bird song, with birds representing spirits of the dead as well as being the vehicle for metaphors about different aspects of Kaluli life. There are different songs for events such as mourning and celebration, based upon different bird morphology and calls. "Becoming a bird is the core

Kaluli aesthetic metaphor because it embodies the emotional state that has the unique power to evoke deep feelings and sentiments of nostalgia, loss, and abandonment."[31] Thus Kaluli sound expressions are embodiments of deeply felt human sentiments.

Musical pieces do not just consist of abstract notes on paper, but those notes have to be interpreted. The same piece of music can be interpreted in various ways according to the emotional, physical, and technical tone that the interpreter gives them. The same could be said in some ways about the psychoanalyst's words. The same interpretation may be spoken in a variety of ways, depending upon the tone required at a particular moment of the analysis. Thus, in the example from Grier that I mentioned, he realised that he could have been critical in the tone of his voice with a patient whose grating high voice got under his skin. However, by examining the musicality of her voice, or rather its lack of modulated musicality, he was able to find somehow a way of speaking to the patient that addressed that discordant part of her, thereby helping her eventually to change the quality both of her voice and of her emotionality. Psychoanalysts thus deal with both language and music simultaneously, and have to come to terms with the complex relationship between them.

Musical listening involves recognition of a complex pattern conveyed by the music, particularly if the listener is schooled in at least some musical analysis, or surprise at a new pattern that emerges from the music, or the music may convey a sense of a *journey*, with the music taking the listener along. That journey may end or keep going, or may taper off into the distance. This parallels in some way the human subject's *psychic* journey and may bring us into the psychoanalyst's territory. The journey may be a long one, as with Wagner, and may involve a trip into the darker recesses of the soul, as with Mahler, or may be brief but encompassing whole areas of human feeling as in a Schubert song, frantic as in a Rossini overture, erotically stirring and exciting as in a pop music concert, and uplifting of the human spirit as when listening to a Bach Passion or Monteverdi's Vespers.

A piece of music, "no matter how short or long it is, can immediately give you the feeling of having lived through a whole life, even if it is a small Chopin waltz, which lasts only, with the 'Minute Waltz', about a minute and a half or so".[32]

Or music can play an integral part of burials, initiation rights, or festivals, often having a precise meaning, whether that be within a religious service or a secular celebration. That is, the meaning of the music is intimately linked to its place in *human relationships*, a theme to which I shall return repeatedly.

There is also something deeply puzzling about how music can have such effects, the nature of musical meaning, and how we can make sense of musical experience. As many thinkers have emphasised, music does not have a clear object (aside from for the listener or a group of listeners) nor an obvious narrative structure; it is not representational in the usual way at least. Clear thoughts are not represented, except of course when words are present as in songs and in opera, though the relationship between the words and the music varies considerably. At times the music and the words go hand in hand as, for example, in Schubert's song cycle *Winterreise* ("Winter Journey"), where each passing emotion of the protagonist has an appropriate musical accompaniment or commentary or addition to clarify his mood.[33] The piano often introduces the musical idea, and then the voice comes in to show how to read the musical idea, focusing our attention on a physical or emotional scenario.

With opera, the music can act as an accompaniment, commentary, or chorus, or to sustain emotion. There may be various kinds of subtle interchange between the words and the music, such as with Janáček's use of Czech speech rhythms to structure his musical organisation.[34] Or the speech may be sung as in recitative between arias and ensembles, or as in *Sprechstimme*, a cross between speaking and singing, the speech pitches may be recited or heightened and lowered according to the composer's directions. George Enescu's opera *Oedipe* uses a kind of recitative style for dialogue, "a vocal style that allows for every emotional inflection to be conveyed, but that can flower into song at moments".[35]

At other times music and words seem to go their separate ways, with different logic. For example, the end of Wagner's *Ring* cycle is one of the most powerful and satisfying experiences in opera. Yet its meaning is also enigmatic. After all the complex and murky plot shifts of *Götterdämmerung*, the ring is finally returned to the Rhine maidens, Brunhilde rides into the fire and Valhalla collapses with the end of the gods. But quite how we get there is not that easy to comprehend and there are many inconsistencies in the plot.[36] The characters are also rather thinly drawn. "Instead of characters in any true sense of the word, these 'personages' are like arenas for the uncontrolled display of emotions that burst in upon their souls from outside, unalleviated by human characteristics."[37] While the mythical scenes seem peripheral to the dramatic structure if considered solely in respect of their contribution to the stage action, "musically they are central: they create the framework for the complex of motivic associations that spreads over the whole drama".[38]

Despite the inconsistencies of the plotting, the music, with its powerful and lyrical "redemption" theme, gives a wonderful sense of a good ending, in which the contradictions seem to have been resolved. Indeed, I think that one can say that in a piece of music like an opera, the music must have primacy for the piece to work satisfactorily. With the ending of *Götterdämmerung*, then, something powerful is represented through the music; it may not be clearly conscious, it may involve contradictions and inconsistencies, but the point is that the lyrical outpouring motive would seem to represent and make tangible powerful *unconscious* urges, yearnings, and desires seeking fulfilment which have motivated the various characters and their emotional outbursts.

Roger Scruton[39] writes that Wagner's music does not merely accompany the singers, nor are they merely singers, but it "fills in the space beneath the revealed emotions with all the ancestral fears and longings of our species, irresistibly transforming these individual passions into symbols of a common destiny that can be sensed but not told". This provides us with a core religious experience for our secular age, through Wagner's use of myth to provide moments of unconscious knowledge, moments that "shine in the darkness of the old Teutonic poems".

One can see already how understanding musical meaning is a complex affair, emerging from interactions at a number of levels. The latter include the composer's intentions and instructions for performance and how they are interpreted by the performers; the space of listening, the concert hall, opera house, church, private space, or arena; the relationship fostered by an artist or conductor with their audience, with their style of playing or shaping the musical texture, ability to communicate the music, and personal qualities; the style of the music, its formal aspects, its dynamics and particular quality of movement, and the historical context of the performance.

The absence of conscious thought presentation in music has led some thinkers to deny that music can represent emotions of any consequence. As I shall explore later in more detail, this implies a narrow view of human emotion, where intellect and thought are separate, whereas most modern research on emotion, including, the latest findings from neuroscience as well as psychoanalytic experience, reveals the intimate link between them.

Indeed, emotions, using a musical metaphor, seem to help us "tune into" the world around us. They are often bound up with strategies for living and relationships with others; they have intentionality and vary from short to long lasting in duration, and very often involve some kind of judgement, however unconscious.[40]

Emotions are thus complex, involving context, meaning, and development, which would seem to make music an ideal medium for emotional communication. "Music is the carrier of the emotion."[41] Music does not signify or merely arouse emotion, instead it "renders emotion tangible, giving a sensuous, reproducible form to something otherwise transient and interior".[42] Music produces an emotional amplification, rendering the expression of the passions more intense and confers on them a heightened energy. "Musical art has the privilege of carrying the emotions to their highest degree—something the most sublime poetry is unable to do."[43]

Thus, from this point of view, the notion of "pure music" untainted by human emotion is unsustainable, using an inaccurate model of an emotion. The complex pattern of a musical work, its context, its weaving in and out of consonance and dissonance, its different rhythms, subjects, and development, all involve complex judgements; without the emotional element integral to the musical judgements, the music would have no life or "soul". And I have already mentioned how Steven Feld's ethnomusical study reveals the intimate relationship between music and emotion.

Indeed, as Dahlhaus shows,[44] the notion of "absolute" instrumental music, purified of its link to text or extra musical references, is a recent historical construct, linked to German Romanticism and the search for the "essence" of things. This contrasts with the long-standing view of music, going back to Plato, that it combines a regular system of tones (*harmonia*), a system of musical time (*rhythmos*), and the expression of human reason (*logos*).

Alternatively, one could see that emotions arise along the circuit leading from the composer to the audience; emotions are not "in" the score, or "in" one element of the chain, but arise as a result of the composer, who often provides marks on the score suggestive of various emotional elements, at least in Romantic music, finding a musical means for the performers to communicate with the audience. But it is finally in the actual *performance* that the emotions are communicated to the audience. One should also add that the satisfaction in appreciating the form of the music, its complex patterns, the realisation of the forms, the to and fro of musical movement, entrances, exits, contrasts, tonal relations in all their dynamic subtlety with heightened tensions and resolutions, actually produces emotions. The experience of listening to music includes wonder and awe, sheer enjoyment, sometimes the physical "tingling" effect or the musical "frisson",[45] which very probably has directly observable neurological causes, but also the intellectual satisfaction of appreciating *form*, which has its own powerful emotional effect on the listener directly linked to eliciting order out of chaos, integration rather than disintegration.[46]

Cross[47] has argued that the very fact that music does not convey specific and singular meanings, what he calls its "floating intentionality", makes it much more suitable for conveying the ambiguity of actual human interactions; it involves a form of communication more adept than language at conveying shared and cooperative interactions.

Theodor Adorno[48] considered the whole dispute about whether music can portray anything definite at all misses the point, and that a closer parallel is with *dreams* "to the form of which, as Romanticism well knew, music is in many ways so close".

He evokes the first movement of Schubert's *Symphony in C Major*, where "at the beginning of the development, we feel for a few moments as if we were at a rustic wedding; an action seems to begin unfolding, but then is gone at once, swept away in the rushing music. Images of the objective world appear in music only in scattered, eccentric flashes, vanishing at once; but they are, in their transience, of *music's essence*. While the music lasts we are *in* it much as we are in a dream."

The sense of touching "another world" like the dream world seems to convey something vital about music, given that we are always going to fall short in language when trying to grasp the nature of the musical experience. As Gustav Mahler put it, "We find ourselves faced with the important question how, and indeed *why* music should be interpreted with words at all ... As long as my experience can be summed up in words, I write no music about it; my need to express myself musically—symphonically—begins at the point where the *dark* feelings hold sway, at the door which leads into the 'other world'—the world in which things are no longer separated by time and space."[49]

The Balinese "speak of the 'other mind' as a state of being that can be reached through dancing and music. They refer to states in which people become keenly aware of the true nature of their being, of the 'other self' within themselves and other human beings, and of their relationship with the world around them."[50]

An ability to imagine "other worlds", as I shall discuss in Chapter Three, requires an ability to go beyond individual existence and represents an advanced theory of mind, first clearly in evidence in cave paintings and in the production of early artefacts, including, at least 40,000 years ago, the first musical instruments.[51]

This other world could be death, or could be the unconscious, the world of dreams. As Martha Nussbaum discusses,[52] the expressive content of music has often been described as having a dreamlike quality, our reactions before it like the experiences we have in dreams, lacking the narrative coherence of our everyday emotions. Mahler describes "this strange reality of visions, which instantly dissolve into the mist like the things that happen in dreams".[53] And music has an affinity with the "amorphous, archaic, and extremely powerful emotional materials of childhood. And it gives them a sharpening, an expressive precision, what Mahler calls a *crystallisation*, that they did not have when covered over by thoughts, in their still archaic form. One enters the 'dark world' in which language and daily structures of time and causality no longer reign supreme; and one finds the music giving form to the dim shapes of that darkness."[54]

Mahler indeed had a rare capacity to evoke childhood reality in musical form, such as with some songs from *Des Knaben Wunderhorn* ("The Youth's Magic Horn"), the *Kindertotenlieder* ("Songs on the Death of Children"), and in the last movement of the *Fourth Symphony*, where, using one of the *Wunderhorn* poems, a child (soprano) sings of a blissful life in heaven. And famously Mahler met Freud for a consultation in 1910 during a summer in Holland (after cancelling three previous appointments), due to marriage problems, which seemed linked to childhood memories of his parents' difficult marriage. Though one must be very cautious about trying to make connections between a composer's life and his music, we have here information from an actual clinical encounter, not wild speculation, which gives insight into aspects of Mahler's musical expression. As Ernest Jones describes in the Freud biography:

> His father, apparently a brutal person, treated his wife very badly, and when Mahler was a young boy there was a specially painful scene between them. It became quite unbearable to the boy, who rushed away from the house. At that moment, however, a hurdy-gurdy in the street

was grinding out the popular Viennese air "Ach, du Lieber Augustin". In Mahler's opinion the conjunction of high tragedy and light amusement was from then on inextricably fixed in his mind, and the one mood inevitably brought the other with it.[55]

Donald Mitchell[56] writes how frequently Mahler's music re-enacted this traumatic childhood experience, "how the vivid contrast between high tragedy and low farce, sublimated, disguised and transfigured though it often was, emerged as a leading principle of his music, a principle almost always ironic in intent and execution". That is, the trauma assumes complex shapes, and becomes an essential element of Mahler's use of musical tension and contrast in his works. In his *First Symphony* this is clear in the way that the slow movement, a sombre funeral march, is based on the round "Frère Jacques", as well as being interrupted by outbreaks of obvious parody, with music that is close to sounding like a hurdy-gurdy. As his music matured, these contrasts became subtler and more integrated into the general structure of the music.

Marcel Proust also connected music with the sense of otherworldliness and of soul meeting soul. Perhaps he alone of writers was capable of grasping the essence of music in words. As he put it in his long novel— during the break in a concert, when people were talking:

> But what was I to make of their words, which like all spoken human words seemed so meaningless in comparison with the heavenly musical phrase that has just been occupying me? I was really like an angel fallen from the delights of Paradise into the most insignificant reality. And just as certain creatures are the last examples of a form of life which nature has abandoned, I wondered whether music were not the sole example of the form which might have served—had language, the forms of words, the possibility of analysing ideas, never been invented—for the communication of souls.[57]

While at least it is generally accepted that music touches the inner depths, there is no agreement about how this occurs and what precisely is touched. Malcolm Budd takes to pieces a number of musical theories, which purport to show how music communicates emotions, though in fact he does agree that music does reach "as far as the inner world of emotion itself".[58] He also quotes Elgar's dedication on the score of his violin concerto, "Herein is enshrined the soul of ..." This secret dedication was likely to be to the soulmate of his later years, Alice Stuart-Wortley. If the soul can be given a voice, then surely this violin concerto would be a leading candidate. As would the plainsong chant of Hildegard of Bingen, *Columba spexit*. The chant creates a sense of sublime serenity and spirituality.

Not only does music have the ability to reach right into the unconscious, but I would also suggest that music, particularly tonal music, resonates at the unconscious level with what I have elsewhere[59] called a *"psychic home"*, the basis of our sense of identity and linked to notions of the *human soul*. The psychic home provides an organising psychic structure for the sense of emerging identity. A psychic home is built up from a number of different elements, as with the physical home, which forms its substrate. There are intrapsychic elements but also intersubjective elements, involving the social world. The sense of home as the ground of our being, the place we need in order to feel secure, is fundamental. This may help to account for the powerful emotional effect of movements to and from a tonal centre, when we hear movements away from and back to a *home* key, in a kind of musical *"journey"*. That journey may have enormous power, as Bostridge[60] points out in Schubert's *Winterreise*, there is a combination of the "homely and the insistently mysterious", perhaps linked to the narrative of an alienated wanderer seeking some kind of home, until he faces the image of death at the end of the cycle.

It may account for the uncanny experience of listening to music such as the prelude to Wagner's *Tristan and Isolde*, where the chromatic tonality remains ambiguous. A diminished seventh chord hovers over the music capable of several different resolutions; we don't know whether or not it will reach home base, creating an intense emotional experience, which no doubt is intended to set the scene for the opera's trajectory of unfulfilled human desire and its deathly transformations.

Daniel Barenboim describes the "psychology of tonality", which parallels the inner life. This is "creating a sense of home, going to an unknown territory, then returning. This is a process of courage and inevitability. There is the affirmation of the key—you want to call it the affirmation of self, the comfort of the known territory—in order to be able to go somewhere totally unknown and have the courage to get lost and, then, find again this famous dominant, in an unexpected way, that leads us back home."[61]

Or as Tom Service puts it, "What are the journeys that your music can take? Either they take you somewhere else and bring you home again, or they take you somewhere else and keep on going."[62]

A main feature of music and a source of its emotional power then is that it usually carries us along; it involves the engaging sense of *movement*, flow, propulsion, or a *journey*. "Forward propulsion is the oxygen of music, the breath of its life."[63] We even talk of "movements" in a piece of music. Music has been described as the embodiment of the physical world in motion,[64] or as embodied expressive movement.[65] The musical journey can touch on deep emotional experiences, with the most profound music there is one might say a journey into the psychic interior, into the human soul, or a *psychic* journey. "In music we sense most directly the inner flow which sustains the psyche, or the soul."[66] One should add that there are also moments when we enjoy the *stillness* evoked by some music, such as a moment of calm in the midst of a stormy piano sonata, or a still centre conveyed by a piece of Tudor choral music.

In a way, music, like the human subject or soul, remains *elusive*. On the one hand, we are enveloped by sound, which has a material basis for its actions, through sound waves. Yet music is a human endeavour, its performance sustained by human relationships; it vanishes into the air once played; it is here and not here in a virtual space of the imagination. It is then both material and immaterial or even spiritual, both real and not real. "In no other human practice does agency depend so specifically on being and not being; in this respect music relates to spirituality."[67]

Yet whatever its nature, movement is integral to the musical journey, while "… every good piece of music must give us a sense of flow—a sense of continuity from first note to last."[68] "In hearing the movement in music we are hearing life—life conscious of itself."[69]

Great conductors seem to have an almost mystical ability to pick out the essential melodic movements in the music and give them concrete shape, allowing us to be taken along by them through the musical journey.

As I shall discuss later, the sense of musical movement may well be linked in some way to the brain's navigational system, the system that helps to organise pace and rhythm as well as physical movement.[70] There is also evidence from developmental research[71] that acquired musical skill and the conventions of musical culture are fundamentally linked with basic bodily rhythms, that there is an *intrinsic motive pulse* (IMP) of walking, marching, skipping, and dancing, which forms the basis for musical activity. IMP is evident in the movements, orientation, attention, and sympathetic expressive responses of infants when they are in musical play with adults, or when they respond positively to fragments of sound. There is also evidence from cross-cultural studies,[72] that music and movement share a dynamic structure that supports the universal expression of emotion, possibly revealing shared brain structures.

In musical terms, one can see musical movement at various levels, from the micro level with relations between pitches to the macro level and the organisation of compositions. The latter is encapsulated by Robert Simpson in his study of Sibelius and Nielsen. He points out that one of the "chief results of the eighteenth-century revolution in the treatment of tonality was a new acceleration in the muscular movement of music".[73] While in Bach a composition swings calmly from harmony to harmony, the music of the early symphonists is more overtly dynamic in the use of key changes. With key modulation, "[t]here is a sense of travelling, and a new kind of musical movement is born." The development of the classical sonata and symphony showed increasing mastery of this kind of movement, until one can see the "special power of movement, the generation of that athletic, organic, and cumulative energy so characteristic of the golden age of symphonic music."[74]

He contrasts this with the mainly serial music of his time, when few practitioners seemed able to produce any sense of movement. There may be agitated rhythms and various chromatic effects, but the music was essentially in his view static. This contrasted with, for example, Nielsen's whole attitude to musical life, which entailed movement as a reflection of life itself. His use of progressive tonality revealed this sense of a "dynamic urge" beneath thematicism, rhythm, timbre, and all the other aspects of music. For Nielsen, experience was movement.

Those composers who still maintained close links with tonality such as Britten and Shostakovich still retained this sense of forward movement in their music. It is interesting that, with minimalism, such as in the work of Steve Reich or Philip Glass, tonality returned to the fore but with a different kind of more gradual movement, akin to ritual.

Traditional tonal organisation[75] was bound up with drama, with shifts between dramatic tension and stability. The dramatic pattern of the music demanded a resolution of rhythmic tension, "a resolution that had to be combined with the need for keeping the piece moving until the end". The core phenomenon of musical movement is thus the tension between tones.[76]

Music is then sound transformed by human experience. Scruton places the sense of movement in the nature of organised sound, in that music or any organised sound is instinctively experienced as an attempt at communication.[77] Thus we hear one pitch like C, and we hear it as a response to the B that preceded it and as though calling in turn for the E that follows. A sense of connection and meaning follows from musical relations.

The conductor Mark Wigglesworth writes that, "Music is about the relationship between the notes, not the notes themselves. It is in this in-between space that expression lies. Between the notes. Between the beats. It is in how you space them apart, how you glue them together, how you connect their consequences or contradict the expectations they create that you metamorphose the printed music into its multi-dimensional expression."[78]

From the point of view of a composer, Adès observes how for him one note can be the fulcral point for the whole piece he is writing. This note can become an obsession around which the piece hinges—a "fetish note". The latter can become a point in the piece where one kind of harmony can transform through a sleight of hand into another sort of harmonic area.[79]

A melody (based originally on identification with the human voice), will then add rhythmic organisation, creating a melodic line, which then takes off, has a beginning and end, as well as a movement between different elements.[80] A melody automatically stirs up emotion, and it brings some kind of sense of narrative; the ups and downs of the pitches appear to be going somewhere, there is a journey and with a journey some kind of story.

One could see here a similarity with psychoanalytic free association, where one thought leads to another and in that movement of thought unconscious meanings, often linked to powerful emotions, emerge. Indeed, the composer Roger Sessions describes the beginning of a musical idea as a "train of thought",[81] its only difference from any other train of thought is its medium being tones, not words or images. One can see this clearly in Beethoven's sketchbooks, where musical motifs are progressively shaped into larger structures. For Sessions, the musical train of thought is largely a "train of impulse or feeling" and it is also a complex mixture of conscious and unconscious processes. Though this is also true of many non-musical trains of thought where ideas and images are highly charged with emotion and that emotion frequently motivates the sequence of ideas, in music it is the logic of sensation and impulse that determines the ultimate validity of the train of thought and gives the musical work not only its expressive power but whatever organic unity it may possess.

Melodic organisation enables a composer to treat a melody or motif as a "subject".[82] This points to the strange phenomenon of musical movement linked to the attraction of sounds to one another, and this gives us the immediate sense of a human context, of a human subject communicating to us: there is meaning not randomness. That is, how the musical trains of thought are put together will be the signature of the composer; there is a *human subject* recognised in the shape of the musical thought; behind the sound is a *musical voice*, which in the major composers at least is recognisable. They speak to us in their own way. They leave their signature on the music; it can even be read in the musical shapes on the score.

Each composer has a particular idiom or style or signature based upon specific musical elements, as well as their own personal development, which establishes their musical "voice". Thus, with Elgar, Jerrold Northrop Moore describes how he forged his style through three terms—rhythmic repetition, melodic sequence, and the subdominant shaping of harmonies, with the frequent use of intervals of fourths.[83] His style's recognisability is based upon the small number of its terms and the concord among those terms. The subdominant impulse gave Elgar's music its most celebrated atmosphere of nostalgia and yearning.

Northrop Moore has also traced a core of Elgar's melodic writing to a tune he wrote when only ten years old, a "Humoreske—a tune from Broadheath—1867".[84] This tune already has Elgar's signature flowing melodic structure. Very much linked to the contours of his childhood landscape in Worcestershire, and in particular at Broadheath, a holiday destination, the tune climbs up and then leads to a long sequence of falling shapes, with a subdominant emphasis. It already reveals Elgar's sense of musical structure and propulsion. "Thus, the child's 'tune from Broadheath' shows the same determined impulse through each of its main constituents: repetition in pulsing rhythm, in melodic shape, in harmonic projection—and in repeating the whole pattern instead of answering it. This first surviving fragment of Elgar's music shows utter unity of impulse and insistence from top to bottom and end to end."[85] It became an essential element of Elgar's emotional language.

One can appreciate the composer's signature through reading their score. With Mahler, apart from the complexity of the writing, he gives in German many specific instructions for the orchestra, much more than is usual. Being a groundbreaking conductor, he was well aware of how orchestras and conductors could stray from the truth of his own intense musical vision, and thus wished no doubt to give clear and detailed guidance to minimise such departures.

With Sibelius the score reveals many sustained brass patterns; his scores, for those who can read music, convey cumulative movement, arising out of slow and swelling musical motion.

Overall, John Blacking describes how every composer has a basic cognitive system that sets its stamp on their major works, regardless of the ensembles for which they were written. "This cognitive system includes all cerebral activity involved in … motor coordination, feelings and cultural experiences, as well as … social, intellectual, and musical activities."[86] This may well include the influence of popular culture on their style, as many composers "have striven to express themselves, and hence their society, in the broadest terms. Lutheran chorales were deliberately derived from folk song, and Bach organized much of his music around them. Haydn, Mozart, and Schubert, in particular, organized their music around the Austrian folk idiom. Bartók, Kodály, Janáček, Copland, and numerous other composers of national 'schools' have found the greatest stimulus in the sounds of their own societies."[87] Dance forms, as well as the melodies of folk music, have also played a vital part in the development and technique of so-called Western art music. One can also add the vital role of subjective inner experiences and the forms and figures of unconscious processes in forming a composer's music, as just touched upon in relation to Mahler.

It is not easy, however, to think about the place of the composer as a subject and their relation to their music. One must be cautious about drawing inferences concerning the relations between a composer's life and their music, unless there is clear evidence, as there is with, say, Mahler or Tchaikovsky, because of the wealth of written material, both letters and conversations. But there should be some way of thinking about the composer as subject of the music, even if that may have little bearing on the facts of their life. There are some artists and writers—perhaps Strindberg is a good example—whose work and life seem inextricably linked. The same perhaps with some composers, such as Shostakovich, even if he were adept at disguising the link—there could be in some of his music an underlying and subversive message against the Soviet authorities. With Nielsen, as I mentioned before, life and music were at one, but with others, the life and the music may seem to be only partially connected; the music may function as a separate realm of creative endeavour or illusion, protected from the turmoil or muddle of daily life.

Edward Cone[88] suggested that the composer makes use of a "persona", or assumes a special role, in order to communicate. While persona and composer as person are linked, the persona also represents what one could see as a special state of mind, or a projection of the composer's musical intelligence, a kind of self-observing ego used in acts of composition. We may never know, or need to know, the actual person of the composer. Yet we do know the composer's persona, what makes their music recognisable, their style or signature, their voice.

In the performance of a vocal piece of music there are a number of different personas or voices—as we have the voice of the composer behind the music, the voice in the music, for example a character, and the actual voice of the performer who themselves will use a special frame of mind to communicate the meaning and emotions of the music, which may or may not be directly related to the emotions of the song. As Ian Bostridge put it:

> The relation between emotion, art and performance is a tricky one. Music undoubtedly arouses emotion, but whether either the composer or the performer has to be in that emotional state to convey or incite it is doubtful. As a performer you can find yourself thinking about the oddest things, fleetingly, while delivering high emotion with a great deal of focus.'[89]

Alfred Brendel[90] writes in commenting on Beethoven's piano sonatas, of the "character" of a piece of music, what in the musical organisation and dynamics gives the piece its coherence, its shape, and its personality. The character can be understood in more or less personal ways, depending on the composition. He points out that even Schoenberg recommended that in composing the smallest exercises, the student should always keep in mind a special character.

Music, like any art form, *distils experience* and goes beyond the merely personal. Artistic, or aesthetic, feelings are heightened and focused feelings. Thus, Beethoven can distil the essence of question and answer in his String Quartet Opus 135 but without words. Equally, Rembrandt can represent and distil the deepest features of human love in his late masterpiece *The Jewish Bride* through painterly means. What music can add, which links of course to the phenomenon of movement, is the *time* dimension.

Different kinds of musical movement and therefore different kinds of music are reflected in different notions of *musical time*. There is the overall historical context for a piece of music, its place in historical time. And here one may have to think about social factors involved in the production and reception of music, as shown for example in Susan McClary's work,[91] which focuses on the way that music produces images of gender, desire, pleasure, and the body. Even the appreciation of a great work such as Beethoven's *Ninth Symphony* is deconstructed by her in terms of the dominance of patriarchal culture. She argues that classical music, no less than pop music, is bound up with issues of gender, construction, and the channelling of desire.

Then there is the linear time frame of a piece of music, the length of the musical journey, a kind of narrative-like structure, with a plot given in terms of the speed of the journey and the staging posts, as well as repetitions and recapitulations; parts of the journey may be gone back over, even if the paths have slightly shifted. But there are more subtle ways of seeing how time comes into the musical structure itself. Thus, Pierre Boulez divides musical time into two forms or poles of a continuum, which are in a constant dialectic in works of music. There is striated or pulsed time (*le temps pulsé*) which is counted out in single or multiple beats, or in two unequal beats in the proportion of two to three—exceptions to these patterns are apparently rare. Smooth or amorphous time (*le temps lisse*) is not related to chronological time except globally; durations appear in the field of time but there is no sense of acceleration or deceleration.[92] Different kinds of music are on different parts of this continuum, with varying amounts of pulsed or smooth time. Thus, in the *Tristan* and *Parsifal* preludes, they both open with an initial slow seemingly endless musical arc, that gradually transforms into a musical pulse.

With the move from Wagner towards Schoenberg, musical expression and the experience of time seems to become increasingly fragmented, less continuous in time and with less of a sense of at least traditional musical movement. This goes in parallel with music's ability to represent complex emotional states.

It is perhaps not too fanciful to compare this duality of pulsed and amorphous time, with Freud's division of the psyche into consciousness which is essentially ordered in linear time, with that of the amorphous and timeless unconscious.[93]

Furthermore, the psychoanalytic notion of deferred action, or *après-coup*, both a translation of Freud's *Nachträglichkeit*, adds to the complexity of how we understand the experience of musical time. *Après-coup* refers to a reordering of experience or a phase of development, after its initial occurrence. Thus, with a young child, once they have acquired language, then past experience can be reordered or reorganised in a new way, after the event as it were. Any past significant event can have new meaning once it is looked at in a different way; there is a reordering of memory and understanding. Applying this to music, one could say that a second hearing of an original theme, its repetition, is never the same as the first hearing, even if the theme is repeated note for note. A second hearing is already a different experience; how we respond to the second hearing will depend upon our experience of the first hearing, and that first hearing will be reorganised in memory by the experience of the second hearing; we hear things again, but that repetition in itself makes us hear what we heard first differently, *après-coup*. This reorganisation in time in itself adds to the sense of musical movement and change.

The role of the psychic home in responding to music may help us to understand the unease, and interest, that we may hear when listening to atonal and serial music, where the sense of home is either absent, disguised, or fragmented; there may not even be any recognisable sense of movement. Not that all music has to convey forward movement. For example, Balinese Gamelan music tends to consist of a continuous flowing line.

Leaving Home was the title of Channel 4's conducted tour of twentieth-century music with Michael Hall and Simon Rattle.[94] "Leaving home" was used as a dominant metaphor for a time when all the social, political, and artistic certainties had migrated. One could also see the metaphor as encapsulating the new approach to musical form, where the old certainties of tonal language were overthrown, at least by modernist composers.

Mahler was already beginning to stretch tonality almost beyond what it could bear.[95] Debussy, in fact influenced by Gamelan music he had heard in Paris, inaugurated the new aesthetic of discontinuity, fragmenting the melodic flow of the music with abrupt changes and frequent interruptions, though still just about within a tonal structure. Thus, by using the whole tone scale in *Prelude a l'après midi d'un faune* (1894), there are no tonics or dominant by definition. The work hovers around the key of E major without ever really getting there, until the very end. Thus, the work does just about reach "home".

Then Schoenberg introduced into the melodic flow a different kind of organisation, where it was fragmented and cut up even more and perhaps closer to the language of the unconscious, thus redrawing the boundaries of the musical journey and also extending the range of human emotions capable of being represented. He talked about "liberating dissonance". One can certainly hear music that seems to take us nearer to extreme or "dissonant" mental states, almost borderline or hysterical at times. For example, in his atonal expressionist piece *Erwärtung* ("Expectation"), a woman anxiously searches for her lover. She eventually finds his dead body in a forest, talks to the body as if it's alive, accusing the lover of betrayal, wonders how she is to live her life now, and wanders off alone into the night. The music has a hysterical quality. Indeed, Theodor Adorno commented on the similarities of this music to a psychoanalytic case study.[96]

Stravinsky had a different solution to the tonal crisis, using free dissonance, with piles of triads used on top of one another as in the *Rite of Spring*, or using polytonality, with the use of more than one key at the same time, as well using new rhythms and melodic fragments.

One could say then that the new music gave expression to the times, where the old uncertainties were challenged, where conscious reason and traditional rationality (represented in music by traditional tonality) were overthrown by Freud's explorations of the unconscious, while new and at times extreme emotional states were capable of being represented musically as a new conception of human subjectivity was taking place. Though serialism dominated much of the post Second World War music scene, in the late twentieth and early twenty-first centuries we have seen a return of tonality in many new guises; it has been given a new home. Musical beauty is now acceptable and even desirable. Not that it ever disappeared; there always have been major composers, such as Britten, Copland, Shostakovich and others, who never abandoned a close relationship with tonality, rather similar to the way that many artists continued to paint figuratively rather than participate in what was the avant-garde abstract mainstream.

In addition, in the 1960s, minimalists such as Steve Reich reconfigured the nature of musical time and experience, introducing new repetitive yet melodic themes and conceptualising music as a "gradual process". Music seemed then close to ritual and to impersonal unconscious forces.

Later Reich wrote: "While performing and listening to gradual musical processes, one can participate in a particular liberating and impersonal kind of ritual. Focused in on the musical process makes possible the shift of attention away from *he* and *she* and *you* and *me* outward toward *it*."[97]

Understanding music and how it affects us is not like understanding a language, though the structure of music has some parallels with language.

Anyone can appreciate a piece of music, as Charles Rosen put it quite simply: "… understanding music does not come from memorising an esoteric code. Many aspects of music … benefit from a long study, but grasping its emotional or dramatic meaning is either immediate or requires only becoming familiar with it."[98] That is, we all have a musical capacity, a mixture of innate and acquired elements. However, he goes on to say that specialised study can allow us to understand why we take pleasure in hearing what we appreciate.

I would suggest that psychoanalysis, as a complex listening discipline, can add something to our understanding of the musical experience at various levels, such as its origins in evolution, its place in the early mother–child relationship and the deep relation between emotion and music, as well as making observations about the complex "listening circuit" leading from the composer and their score to the performer and their audience. Musical studies in turn can enrich and clarify aspects of the analyst's listening function, as I have already sketched out in the work of Lombardi, Reik, Grier, and Nagel. Analytical and musical listening share similar states of mind, experiences, and cognitive processes, in particular in the use of non-verbal communication of emotional states and unconscious meanings. I shall endeavour to explore the various ways that music and psychoanalysis can move across their separate disciplines and communicate with one another, starting with early musical experiences, the root of music's power to move us emotionally.

CHAPTER TWO

Early musical experience as a root of music's power

Because music occurs in time, it can under certain circumstances provide a powerful sense of *continuity*. Already with the early mother–baby relationship one can see how the *maternal voice* echoes and re-echoes to the baby's sounds, in a kind of musical manner, imitating and repeating what comes from the baby and providing, as Daniel Anzieu[1] describes, a sort of *sound mirror*, not a static mirror but a dynamic and responsive mirror providing a sense of continuity over time. Thus:

> Winnicott … lists babbling among the transitional phenomena, putting it on a par with other activities of this type. But a baby will only stimulate itself by making sounds while listening to them if it has been prepared by an environment that has immersed it early enough in a bath of sounds of the right quality and volume. Before the gaze and smile of the mother who feeds and cares for it reflect back to the child an image of itself that it can perceive visually and internalise in order to reinforce its self and begin to develop its Ego, the bath of melody (the mother's voice, the songs she sings, the music she lets it hear) offers it a first mirror of sound, which it exploits first by cries—which the mother's voice reacts to with soothing noises—then by gurgles, and finally by playing with phonemic articulation.'[2]

The space of sound, or what Anzieu calls an *"audio-phonic skin"*, or what Guy Rosolato also called a *"sonorous envelope"*[3] is the earliest psychical space, beginning within the uterus. All sorts of sounds face the developing baby, noises from outside which can cause pain and distress when loud or sudden, strange gurgles and noises from within the baby's body that can be alarming, the surrounding world of adult speech, sometimes booming, sometimes calming. In turn, the baby's cries, from hunger, pain, or frustration both elicits a response from the caregiver, and can be a further source of alarm to the baby. The mother's soothing voice along with gentle movements is clearly the most effective way of calming a young baby, and babies are able to discriminate this voice from others very early on. I think it important to add, following Winnicott, that this early sonic space is one built up by *both* the baby and the mother: it is a mutually created space, the first intersubjective space.

I don't think we emphasise enough that the analytic encounter takes place within a complex, boundaried *sound world* or *soundscape*, in which hearing takes precedence over seeing. Freud's use of the couch was a radical way of pushing the sound world to the fore in treatment. The distribution in space of things heard is fundamentally different from that of things seen.[4] Sight tends to distance us from things; there is a landscape which we can admire but it remains out there. Figures may move in a landscape but the landscape we see does not move. But hearing envelops us. "Sound, by its enveloping character, brings us closer to everything alive."[5]

To see without hearing is to witness an uncanny dumb show, and is disorienting. But to hear without seeing, as in closing one's eyes, can be revelatory.[6] One gets more profoundly in touch with moods, emotions, and the meaning of words.

"Hearing *musical* sound, with or without words, makes us especially aware of proximity and thus connectedness. Parents sing lullabies to their infants, and their infants respond: this is music at its most enveloping."[7]

However, as Anzieu describes, the sound mirror may become pathogenic, when for example it becomes dissonant, as when the mother intervenes in ways that contradict what the baby is feeling. Or it may become abrupt, with the mother's sudden shifts of mood and attitude, confusing the baby. Or it may be impersonal when the mother, either through depression or due to an emotionally cut-off personality, reacts as if the baby is a machine to be programmed rather than responded to. The mother may talk in front of the baby but not about the baby, who is basically ignored. Sounds then become for the baby a signal of rejection or of a deadened emotional world, and the mirror of sounds fails to provide positive emotional information for the baby. Otherwise, the mother's vocal responses normally provide a positive experience for the baby, enveloping or wrapping the baby in a comforting and enlivening sound world.

In distorted mother–baby relationships, for example with a depressed or borderline mother, there may be a lack of responsiveness, and the maternal echo can become more like the plaintive echo in the myth of Narcissus, and time can become deadly, what Green has called "dead time".[8] The depressed mother's speech is less musically expressive and less focused on the infant.[9] Borderline mothers tend to be unpredictably intrusive or withdrawn and express more negative affect; their vocal performance is also inconsistent and dissonant, reflecting their shifting mood states, and that is confusing for the baby.

The maternal voice has also become emblematic of fantasies of union with the early mother, prior to speech and language.[10] And of course the female voice has all sorts of powerful culture associations, both negative and positive, from the image of the siren luring men onto the rocks to their destruction in Homer's *Odyssey*, to the various enchantresses of opera, seducing, enticing but also exciting men, and indeed women, from Poppea to Carmen.[11]

The theme of women finding their voice in a predominantly male culture, where their voice has often been silenced, or has had to go underground, has powerful resonances. Thus, as Leslie Dunn and Nancy Jones write in a collection of papers devoted to female vocality in Western culture, "Feminists have used the word 'voice' to refer to a wide range of aspirations: cultural agency, political enfranchisement, sexual autonomy and expressive freedom, all of which have been historically denied to women."[12] However important these issues are for women wishing to find their own voice and sense of authority, there is also, as Dunn and Jones emphasise, a need to return to the nature of the literal audible voice in all its complexity, upon which the metaphor was based. Indeed, the conductor Mark Wigglesworth writes that the human voice's ability to move us "is so strong that almost all music aspires towards a vocal quality. Conductors are always asking orchestras to make their instruments 'sing', whatever form of music they are playing."[13]

But with regard to the mother–infant relationship, Marie-Cecile Bertau[14] has described, from a psycholinguistic perspective, how mutual imitation of child and caregiver in early communication and speech acquisition form an incessant movement from one to the other, intermingling the voices of both, and becoming the basis of internalisation. As she emphasises, *voice* plays an important part in raising children. It accompanies, structures, and creates a sense of rhythm during the physical aspects of handling the child. "The voice stresses a certain quality of the caregiver's action: slow, smooth, rapid, impatient, etc."[15]

Almost all cultures have developed a form of baby talk, or what has been called "motherese", or "infant directed speech" (IDS) where besides semantic and syntactical features that reduce complexity, it is the voice quality of the caregiver that matters; it is of a higher fundamental frequency, slower and with clear intonation, and with a lot of repetition. IDS itself varies, depending on the baby's age. To begin with it serves to engage and maintain the baby's attention, then it helps to modulate arousal and emotion, and then it gradually communicates the parent's own feelings and intentions, helping to shape the mother–baby relationship. These kinds of exchange appear to be present in all cultures, providing some evidence of the universality of musicality and its key role in the regulation of emotion.

One could add that these early pre- and non-verbal exchanges between the caregiver and the young child provide the scaffolding for subsequent speech development and social interaction. Language does not simply replace preverbal forms of communication. The child, though not able to speak yet, still acquires the fundamental interpersonal skills necessary for social life, through observation, receiving physical holding, and the intense sound communications between mother and child. In these early years, babies and young children are learning about the basic processes and structures of interpersonal exchange, the flow of feeling, and the shape of emotional life that underpins social communication; without these early "musical" elements, without pitch variations, and without social connectedness, speech and language remain wooden.

The child is in the realm of what Julia Kristeva calls the "semiotic", where the baby, not yet a speaking subject, still makes sense with intonations, vocalisations, and echolalia; this is prior to entering into the symbolic world of language or of "signification", when the baby can use phonemes, morphemes, and syntax. She points out that a tonal language like Chinese retains this semiotic, musical capacity and integrates it into the symbolic system, making it signifying in linguistic communication.[16]

A further complication in understanding these early mother–infant exchanges is that language is present during the interactions, but on the mother's side. She will often be putting things into words, through a kind of singing, partly for the baby and partly for herself, in order to provide a context for communication. "There you are, aren't you lovely? Oops, what have we here? Is it time for a nappy change? Where's Daddy, where's your little sister? Ooh, here we go, up we go, that's the way," and so on. There is thus a mismatch between baby and mother, but that reflects the extreme helplessness and dependency of the baby on the mother.

The French psychoanalyst Jean Laplanche, with his theory of seduction, has a very different take on the early mother–infant relationship, less based upon empirical research and more on psychoanalytic theory and adult analytic practice. He sees this early situation as one involving the adult providing a sort of sexual seduction of the infant, given the disparity between the adult world of language and the infant's immaturity. The adult provides so-called "enigmatic signifiers" which are experienced by infants at some level as traumatic, as they cannot process them; they have yet to have an unconscious, and so are invaded by these enigmatic messages from the mother; they are invaded by her sexuality. This inevitable traumatic situation will mark the child's development and becomes foundational for the construction of the unconscious.[17] Though this is a theory with little basis in actual knowledge or observation of babies and is, rather, a projection backwards from the adult onto the child world, it does point to a basic dilemma that there is a mismatch between adult and child worlds that both sides have to negotiate.

There is also the role of the many different voices that the child will encounter, both from the mother herself and from others in the environment, out of which a sense of self based upon dialogue will hopefully develop. Bertau emphasises the voice as an inherently dialogical phenomenon, providing the basis for an alive social life. Her emphasis comes from her reading of the Russian school of literary theorists and psychologists such as Mikhail Bakhtin and Lev Vygotsky. Thus, in his study of Dostoevsky, Bakhtin took the view that what represents Dostoevsky's unique and revolutionary style is that there exists in his narrative a "plurality of independent and unmerged voices and consciousnesses, a genuine polyphony of fully valid voices".[18] Bakhtin argued that these novels are "dialogic", and use a pluralistic framework, which he contrasts with the traditional "monologic" form of novel with the single and omnipotent authorial consciousness.

Bakhtin extends this literary reading to the nature of dialogic thinking itself which, arising out of human dialogue and encounter, is more able to discover truth than monologic thinking, which possesses ready-made truths. One can see how Bakhtin's view of communication involves communion between what I have called the "many voices of consciousness", both within the individual and between individuals.'[19] Consciousness has many streams or voices. Through dialogue the themes of these voices may become more or less coherent. In the analytic encounter, one could say that the patient may move from monologue to dialogue, from having a single or reduced stream of consciousness, with one or few voices, to the capacity to experience many voices, allowing them, as in a sort of fugue or symphony, to penetrate and overlap. The early mother–baby exchanges provide the framework for alive and "tuneful" dialogue, or on the contrary the first elements of what may become pathological or lifeless monologue.

Vygotsky similarly emphasised the social origins of mental life, the latter requiring interaction between people. The internalisation of psychological functions such as memory, language, and thought requires a social relationship. Bertau uses this notion to describe internalisation as a result of the child experiencing the voice of a significant other. "The specific intonations and the expressive, idiosyncratic style of the person as manifested in her voice give a specific taste as to what is internalized".[20]

Maya Gratier and Colwyn Trevarthen responded to Bertau's essay by proposing that "Voice cannot be separated from the stream of moving in time, from the 'musicality' of it; the rhythms and cadences of its expression and in memory of being in company."[21]

Based on years of mother–infant research, they describe how the "behaviour of infants in their delicately negotiated engagements with sympathetic partners and playmates demonstrates that there is an *innate intersubjectivity* that enables synchrony of intentional rhythms, expressive gestural forms and qualities of voice with others from birth".[22]

Indeed, there is good evidence that human infants are well equipped to learn the musical regularities of their environment. Thus, at least by six months, babies have the probably unique human ability to recognise relative pitch perception, the ability to recognise that a melody is the same when the pitches are transposed up or down. By one year, the child can show sensitivity to musical keys.[23] Furthermore, babies can detect small differences in speech sounds from all the languages in the world. You can test this by how hard a very young baby sucks on a dummy attached to a computer, or a bit later how babies learn to turn their head to a new sound. Babies are born with this ability to discriminate sounds, but then lose it from around nine months, when they learn to home in on the speech sounds of their native language (or languages if they are brought up in a bi- or multilingual home).

Babies are also more sensitive to aspects of musical rhythm than adults, and can detect subtle and complex changes in rhythmic patterning. In addition, they can detect one note difference in the key of a melody that adults completely miss. As with detecting speech sounds, babies lose this ability at about a year old when their perceptions narrow. While the ability to move in synchrony to a musical beat requires complex motor skills and does not appear until about five years, babies do move rhythmically much more to music than to speech.[24] This research has shown that infants engage in significantly more rhythmic movement to music and other rhythmically regular sounds than to speech; they exhibit tempo flexibility to some extent: for example, faster auditory tempo is associated with faster movement tempo; and the degree of rhythmic coordination with music is positively related to displays of positive affect. These findings are suggestive of a predisposition for rhythmic movement in response to music and other metrically regular sounds. Not surprisingly, singing is more effective than speech in calming infants.[25]

Gratier and Trevarthen, using for example detailed study of filmed mother–baby interactions, describe how by six weeks infants use their voices, and the expressions and gestures of their body, with powerful conversational intent; their vocalisations are invitations to the caregiver to engage in repartee with them. "Protoconversations" are already going on by two months, with the baby using expressions of eyes, face, voice, and hands which in turn triggers the response of the parents' active encouragement and engagement. Parent and baby soon move in synchrony to the same tempo and dynamic shape; by three months one can already observe playful interactions between parent and baby. Thus, one can see early on that intimate verbal exchanges between the mother and baby are already constructing a proto-narrative.

Stephen Malloch[26] used computer-aided musical acoustic techniques to study vocal exchanges between infants and adults to clarify how the pulse and expressive and emotional qualities of voices are engaged in improvised "musicality", or what he called "*communicative musicality*".

Malloch is using a special and wide use of the term musicality here. He is referring to innate human abilities, across cultures, that make music production and appreciation possible, involving an innate skill for moving, remembering, and planning in sympathy with others, not only as seen between mother and babies, but any human situation involving close and collaborative interpersonal communication.

The three elements of communicative musicality as vital for sympathetic parent–infant communication consist of:

1. *pulse* and the examination of timing of responses,
2. *quality*, with the pitch contour of vocalisations and their timbre, the latter consisting of the three measures of roughness, width, and sharpness, and
3. *narrative*, which combines pulse and quality, showing how the mother–infant pair share a sense of passing time as the mother, for example, chants a nursery rhyme.

Furthermore, Trevarthen[27] postulated that humans move under the coordinated and integrated control of a time-keeping energy-regulating "intrinsic motive pulse" (IMP). While skilled auditory processing of musical information and the cultivated appreciation of music requires many higher brain functions, the basic IMP originates in the brainstem. The IMP is conceived as a series of generators of neural and body-moving time, which forms part of a larger system of generators regulating our emotions, movements, and thoughts. As I mentioned in Chapter One, there is evidence from this developmental research that acquired musical skill and the conventions of musical culture are fundamentally linked with such basic bodily rhythms, that the IMP of walking, marching, skipping, and dancing forms the basis for musical activity. IMP is evident in the movements, orientation, attention, and sympathetic expressive responses of infants when they are in musical play with adults, or when they respond positively to fragments of sound.

Gratier and Trevarthen[28] propose that the expressive rhythm of human voices, or the communicative musicality of the mother–baby interchange has a vital role in promoting the well-being and comfort of the baby. Using Winnicott's notion of physical and mental holding, they propose that the vocal rhythms of interpersonal engagement constitute a holding environment for the infant that is in continuity and coherence with the physical holding involved in the caregiver's mothering techniques.

They postulate that infants who are securely attached are those who have received a continuous and coherent "musical holding". Holding then consists not only of consistent handling of routines but also in tandem with consistent and well-attuned rhythmic patterns, with stable and predictable vocal pitch, contouring, and timbre.

Thus, overall, one can say that communicative musicality is a vital element of the bonding and attuned attachment between mother and infant. Without musicality the internalisation from the interaction between voices is distorted and emotions are disturbed. As I shall discuss in the next chapter, one could well imagine that there was a strong evolutionary advantage in developing communicative musicality, given it is such a crucial element of mother–infant bonding.

Gratier and Trevarthen in addition make the point that a mother's voice is also the voice of her community; it carries the imprint of others' styles of speech. In that sense the baby is early on exposed to the conventional styles of the community, with all their particular speech rhythms and tones, forming the basis for *belonging*, for feeling at home in a safe and caring community, or what one could call a "*speech home*", one of the dynamic elements of the psychic home.

Daniel Stern[29] had described the important role of "*vitality affects*" in the mother–baby relationship. This was a way of trying to describe the dynamic quality of the emotions between mother and child, and how a mother may be "tuned" into the baby's state of mind or on the contrary have difficulties in so doing. *Affect attunement* is an important quality in good enough mother–child relationships, and something that needs to be looked at when considering the nature of attachments. He cited the work of Suzanne Langer[30] who had already paid attention to the many "forms of feeling" inextricably involved with the vital life processes. She had also used the notion of forms of feeling to capture the many feelings evoked by music. For her music does not so much evoke particular feelings but their "form", their essential shape over time.

In his later work, Stern[31] extended the notion of vitality affect and, rather in the manner of Langer, described the role of "dynamic forms of vitality", a mental creation shaping human experience, including the musical experience.

Music, one could say, allows the adult to get in touch with intense early experiences related, however indirectly, to the vital early mother–child interactions, giving both pleasure and release of emotion; hence one important source of its power to move us.

Vitality forms can be described in terms of movement, time, force, space, and direction, all together giving the experience of vitality.[32] Dynamic forms of vitality give life and shape to the narratives we create about our lives. We tend to think of the mother–baby interaction in terms of objects and space; the advantage of this way of thinking is that one is dealing with the "real time" phenomena of process, dynamics, and flow.

Movement is essential to dynamic forms of vitality. When a mother's face stops moving, the baby becomes upset very quickly. Without motion we cannot read or imagine mental activity, emotions, and thoughts in a person we meet. An embodied mind is one that reveals a temporal shape in action. The vitality form gives temporal shape and force to an emotion, which in turn provides the direction, goal, felt quality, and tendency to act in an interpersonal situation. Affect attunement can now be seen as based upon the matching and sharing of dynamic forms of vitality; its "frequent use permits a mother to create a degree of intersubjectivity higher than faithful imitation".[33]

Dynamic forms of vitality are part of episodic memories, memories of specific events, situations, and experiences, and give life to the narratives we create about our lives. For example, thoughts and feelings may suddenly surge up, fade in and out, as when one recalls a past event, or may be evoked by a present perception that in turn conjures up past events, just as Marcel Proust described when his tasting of the madeleine cake conjured up past associations of his early life.[34]

With regard to music, Stern proposes a parallel between the shape of the vitality form, its patterns of ups and downs, shades of intensity, and flow, with the musical phrase and its dynamics. When we hear music, we experience sound in motion. Dynamic forms of a piece of music are written into the musical score. The difference between an adequate and moving performance lies in the "unique vitality dynamics that a great artist can bring to the work and transmit to an audience".[35]

Stern considered that communicative musicality as described by Malloch and Trevarthen is largely based on the coupling of vitality dynamics between people. "Musicality' is composed of pulses that are formed by timing, in the rhythmic sense, and its temporal contouring, and the deployment of force in time. This is the backbone of vitality dynamics where 'being with another' is accomplished by sharing the vitality dynamic flow."[36] Hence one again can see the natural links between emotions and music, a theme to which I shall return repeatedly.

Stern also touches upon how such understanding can shape even such sophisticated musical products as Beethoven's *Fifth Symphony*.

For example, the famous opening four pitches establish an initial level of arousal and a specific vitality form. Then they form the basis for many variations in intensity, speed, timbre, colour, and stress, shaping the pattern of arousal up and down. "Beethoven's subject matter is nothing less than the vitality dynamics of music and life."[37]

While one may be sceptical about how far one can apply the notion of vitality forms to a sophisticated piece of symphonic writing, it does seem to add something in how we can appreciate how emotions can be communicated by music, and it does add a dynamic and temporal account of music, which more purely structural analyses can leave out.[38] Furthermore, the dynamic features of a piece of music are key to being able to recognise what kind of music one is listening to, be it rock, blues, reggae, the list goes on, and what to listen for. The dynamic markers on the musical score, such as loudness and softness, changes in intensity, stress, repetitions, and so on, are key to the way that "rhythms and melodies with their variations and harmonies are imbued with the feel of being alive and vital".[39]

One could also compare the function in Wagner's operas of various motivic themes, the so-called leitmotifs, to the role of vitality forms. These motifs vary from links to particular people or their objects or activities; they could represent fate, the sword, Valhalla, Siegfried's horn call, and so on. They consist of melodic, harmonic, and rhythmic cells which can be transformed and combined in various ways to match the progress of the drama. Their changing and shifting contours seem analogous to the way that vitality forms can shift and change.[40]

Emotional connectedness and attunement to one another are essential elements in musical performance, perhaps owing something to these early musical foundations of intersubjectivity. A combination of cooperation and individual expression takes place, with the musicians mutually adjusting their actions and sounds, particularly in small ensembles, where close mutual and intersubjective and empathic listening is essential. This is a form of "reflecting-in-action"[41] where one adjusts as one is performing. It is most clearly discernible in the improvisations of jazz musicians, who reflect through a "feel" for the music. But it is also clearly *visible* in any small ensemble; one can observe the human context, the intense communication between the players, which allows for an intense musical experience. I think the visual experience for the audience is part of the pleasure of experiencing the music. How a pianist for example comes on stage, engages the audience with the eyes, hands, and gestures is all part of the aesthetic experience.

Some orchestras such as the Berlin Philharmonic are able to arrange for their players to be able to play a fair amount of chamber music outside their orchestral duties. This makes for a special capacity to work together and therefore an especially rich sound. A key notion in this context is that of the synchronisation of the body with the environment or "*entrainment*", that is, the "alignment or integration of bodily features with some recurrent features in the environment".[42] Musical entrainment involves perceiving the regularity of beat and can be seen for example when dancing to music or marching in time to music. It seems, as I shall discuss later, to be hard-wired into the motor centres of the brain, since it is a skill that children can be seen to acquire naturally. There is even evidence that participating in musical activity such as synchronised singing and drumming can promote cooperation in four year olds.[43] This research shows that joint music making among four-year-old children increases subsequent spontaneous cooperative and helpful behaviour, relative to a carefully matched control condition with the same level of social and linguistic interaction but no music. It seems that music making, including joint singing and dancing, encourages the participants to keep their eyes and ears in touch with the group sense of togetherness, and enables them to sing and move together in time—thereby effectively satisfying the intrinsic human desire to share emotions, experiences, and activities with others. Such collective power can be seen in many social situations, including music's role in facilitating group *identity* in young people, where music can create a collective mindset, be a focus of a way of life, a style of being, including what to wear and how to speak, facilitating consensus or subversion.[44]

With musicians there is obviously a complex form of entrainment, where they "regulate the temporal alignment of their musical behaviours by engaging in continual processes of mutual adjustment of the timing of actions and sounds".[45] Ensemble performance involves acting in synchrony and requires "constant temporal negotiations ... that are at once cognitive and embodied".[46] This involves conscious and unconscious communications between the players, communication at both the bodily and emotional level, with the reading of gesture and eyes as well as the building up of trust and mutual understanding. Emotional focus, where the performers are enabled to be absorbed and focused *within* the music somehow seems to be a vital part of giving a good performance,[47] and requires this sort of close common understanding and communication.

Perhaps we can understand some aspects of the psychoanalytic relationship in these terms, where there may be different degrees of entrainment between analyst and patient, depending upon the nature of what gets repeated in the transference.

With orchestral playing, the complex role of the conductor adds further to the need for musical entrainment. Gone are the days of the conductor as dictator, instilling fear and using and abusing power, or even those who wish to impose their will by empathy with the minds of the players, as Bruno Walter described.[48] Instead, conducting is really about facilitation and mutual music making as well as collaborative leadership. Tom Service in his intimate study of a series of great conductors such as Claudio Abbado, Simon Rattle, and Mariss Jansons, describes their work as, "the creation of a culture of responsibility, of respect, of musical and social awareness, and of listening ... Through the creation of a virtuous circle of listening, which starts with the conductor working on the piece when they are preparing it, continues in the enlarged space of the rehearsal room, and finds the outermost of these concentric circles in the inclusion of the audience in the final performance, conducting is a metaphor not for absolute power but for shared experience, for collaboration, for listening."[49]

Daniel Barenboim describes how in musical performance, "[T]wo voices are in dialogue simultaneously, each one expressing itself to the fullest, whilst at the same time listening to the other."[50] This kind of communication, obviously close to the kind of intersubjective communication I have already referred to, is not solely about music but is a lifelong process. For Barenboim this capacity of music for engaged conversation, its dialogic quality, can help in mutual understanding between people who might otherwise be deaf to what they have in common. His West–Eastern Divan Orchestra, formed from Israelis, Palestinians, and Arabs is a concrete manifestation of this hopeful principle. One can see here the power of music's ability to bring people together in a mutually satisfying endeavour, breaking down barriers to understanding and facilitating and heightening mutual communication.

The way that the conductor's elaborate gestures conveys meaning to the orchestra varies a great deal. Thus, Valery Gergiev

> doesn't bring music into being with upbeats or down beats, but with tremors and vibrations of his fingers, with explosive propulsion system of his elbows, with violent convulsions of his shoulders, and even with

bestial laryngeal grunts. There is no perfectly observed rhombus or triangle anywhere in his technical or physical apparatus. Instead, there is a tremulating mass of nervous energy that infects the air around his hands and eyes and which somehow communicates something to the musicians.[51]

Jonathan Nott makes the point that being too correct in emphasising each beat can have the paradoxical effect of making the players play less well together: "It's better to be less correct and more vague in your gestures, so that the music moves between the beats instead of being limited by them."[52] Wagner's book *On Conducting* in 1869 already made the point that the tempo indicated by the conductor should be imbued with life and will vary according to the needs of the music; the conductor will find the right comprehension of the "Melos" or the melody.[53]

Lorin Maazel is the opposite, giving precise indications for every shade of musical expression from the orchestra. Whereas Simon Rattle recommends that "You have to learn to allow them to listen, and that means being less rather than more precise in your gestures."[54]

Thus, the conductor's gestures are not just to indicate the beat but also to create a *frame* for close communication with the orchestra. Most modern conductors now seem to see conducting as allowing things to happen by enabling the members of the orchestra to listen closely to one another as they respond to the music. This is indeed very sophisticated musical entrainment, based upon experience and assumed technical mastery as well as close reading of the composer's score, requiring complex interpretative skills.

Mark Wigglesworth makes the point that physicalising emotion is something people do every day, so that conductors "simply take a basic fact of life and use it as a means of creating a clarity of musical style and a strength of emotional feeling."[55] A conductor has to feel the sound in the body, which in some ways simplifies the task of communicating with the orchestra. Words may be used in rehearsals to make a musical point, but cannot be used in the performance. The conductor's body has to convey the quality of the music's pulse, rhythm, movement, and the musical logic and shape of the music, all with the right emotional charge that can both lead and inspire the performers.

While such a task requires insight into the truth expressed in the musical score and sophisticated musical knowledge, in order to provide the convincing movements that capture the shape of the musical phrasing, the conductor's means for carrying out this task has close links with the early communicative musicality. Indeed, this is recognised by Wigglesworth when he writes that:

> A conductor paints emotion in the air in exactly the same way as someone who is not a conductor would do it. You do not have to be a mime artist to express a wide range of things though gesture alone. An elemental form of human communication, learned from birth, is instinctively expressed and instinctively understood. In that sense alone, conducting is simply child's play.[56]

In order to create a powerfully convincing performance, a conductor needs to be able to *listen* to how the "musicians respond to their beat in order to know how to conduct the next beat. There is no contradiction between listening and leading. It is a constant cycle of action and reaction, a virtuous cycle of coordinated response that connects the reality of the sound with the sound of your imagination."[57] A listening dynamic is set up between orchestra and conductor, and at times the best way for this to be achieved is for the conductor to do less rather than more, to be still in order to encourage both musicians and audience to travel inward, into their deeper selves, allowing the music to "foster an invisible coming together of everyone in the room."[58]

Such considerations emphasise that, as Christopher Small has described,[59] music is not a thing but an activity, something that people do, often together, and that this involves human *relationships* in all their richness and complexity. It was Small, I believe, who first coined the term "musicking", the present participle of the verb to music. "To music is to take part, in any capacity, in a musical performance, whether by performing, by listening, by rehearsing or practising, by providing material for performance (what is called composing), or by dancing."[60]

Musicking involves human encounters taking place in a specific setting. For Small, the act of musicking establishes in the place where it happens a set of relationships, and it is in those relationships that the meaning of musicking lies, whether or not they occur in a concert house or in any community setting. Musicking then is an activity "by means of which we bring into existence a set of relationships that model the relationships of our world, not as they are but as we would like them to be, and if through musicking we learn about and explore those relationships, we affirm them to ourselves and anyone else who may be paying attention, and we celebrate them, then musicking is in fact a way of knowing our world … and in knowing it, we learn how to live well in it."[61]

He adds that in a sense the sound relations of a musical performance stand in metaphorical form for ideal human relationships: presumably one of the reasons why musical performances can be so emotionally powerful, another theme to which I will return when considering more specifically the relations between emotion and music.

The processes of emotional and musical entrainment I have described, with music's power to synchronise emotions and actions, seem to have their origins not only in early communicative musicality but may go back some way in evolution.

I have already referred to Ian Cross' account of music as capable of communicating ambiguity and of being a form of communication more adept than language at conveying shared and cooperative interactions. From this follows his argument that the faculty for music is as a result likely to have had strong *evolutionary advantages* for humans in their interactions, as covered in the next chapter.

CHAPTER THREE

Origins of music's power

Music's origins in evolution

Given that music plays such an important role in many people's lives throughout the world, and the good developmental evidence that our early lives are permeated with a form of musical communication, and that at least a certain amount of musicality is innate, or at any rate there is good evidence that human infants are well equipped to learn the musical regularities of their environment,[1] it is not surprising that there has been considerable interest in, and speculation about, when and how musical activity, or musicking, emerged in human evolution.

The oldest surviving musical instruments are musical pipes about 40,000 years old, from various sites in south-west Germany. Two complete pipes come from Geissenklösterle and are made from mammoth ivory, and another complete pipe, made from the wing bone of a griffon vulture, comes from Hohle Fels. They are all of a similar design with finger holes arranged along the pipe, which could thus produce discrete pitches.

The pipe from Hohle Fels seems to have been made by scraping across the tube so that the delicate structure would not fragment. That from mammoth ivory needed more complex construction; it required hollowing out of a broken bit of ivory and then piecing bits together with some sort of adhesive.[2] New techniques were not needed to make these pipes, as by then humans had been using these sorts of techniques for making various tools for thousands of years. It is likely that by the time these pipes were made, humans were already making music, either with these kinds of pipes or with percussion instruments. But it would be clear enough that by about 45,000 years ago, musical production had become a feature of human culture, and its development would then be based on cultural changes. One must at the moment speculate about how humans and their ancestors got to this point.

However, organised cave paintings date from around 25,000 years ago, though various organised marks are traceable to around 40,000 years, as are some animal figurines found in southern Germany. Recent discoveries from various parts of the world have unearthed perforated shells which appear made for stringing and go back around 100,000 years, while various beads and other worked-on shells can be traced back between 40,000 and 70,000 years. Thus, there is reasonable evidence of the existence of human symbolic activity going back to 100,000 years ago. By then one could probably assume that humans were communicating with one another, had some sense of "other worlds" beyond their own horizon, and that this would involve some form of language, though the first writing we know of was not until about 5,000 years ago and the first numbers a little before that. Quite where music came in this story prior to 40,000 years ago is difficult to know, but the same problem arises when speculating about the origin of language. At least we have a clear idea, through the ancient flutes, when humans were playing music; there is not such concrete evidence about when humans were speaking to one another save for the reasonable deduction from the survival of symbolic artefacts such as paintings, sculptures, and organised burial sites.

It is worth adding here that though language and music both involve organisation, structure, sound, and human communication, they are essentially different phenomena. There have been various attempts to use linguistic concepts and methods as a way of explaining music's organisation, such as by Fred Lerdahl and Ray Jackendoff with their "Generative Theory of Tonal Music",[3] but the general view now is that, though music has a structure, this is different from that of language; music has no semantic structure, and though it is a system of organising sounds this is different in kind from how language organises relationships between words. Nonetheless the analogy between music and language can be quite useful, as long as its limitations are recognised.[4] I shall return to the usefulness of the analogy when considering once more the relationship between music and emotion, in the next chapter.

Psychoanalysts are often in a position to wonder about the origin of their adult patients' psychological difficulties in a way which has some parallels with this issue of the origin of music and of language. Looking for a single answer is usually a waste of time; instead symptoms and personality features are overdetermined, as the result of multiple factors in a complex web of past and present relationships. The same is likely to be true when wondering about when and how human language and music emerged.

There are various possible ways that this emergence took place. Thus, music and language might have evolved together or separately. Music may have evolved prior to language, as today musicality in babies can be seen prior to their use of speech; or language may have arisen prior to music, so that music "leant on" the language system for its development and expression. Music and language may instead have evolved in parallel, with similar but essentially separate lines of evolution, each answering different human needs, and preceded by patterns of sociality and communication neither musical nor linguistic, but then coalesced at some point late in human evolution, as suggested by Gary Tomlinson.[5] Music may have enhanced language evolution by providing an increased emotional element to communication, such as in song, and may have been a kind of basic proto-language of emotion. Music may have been a mere by-product of language evolution, the suggestion being that it was language that was an evolutionary adaptation, enhancing memory, attention, categorisation, and decision-making, while music was just a chance "add on", its main purpose being that of giving pleasure—the controversial suggestion of Steven Pinker.[6] There might have been a common "musi-language" out of which music and then language separated, which is Steven Brown's preferred option.[7] Or as Steven Mithen[8] has suggested, some form of primitive communication system formed the basis of the subsequent evolution of both music and language.

Of course, it may just be the case that early humans found themselves making noises out of flutes rather by chance, but then rather enjoyed the effect. They then played with different shapes and thereby made different tunes. They began to play with the flutes and to enjoy playing around and that made them feel good. Maybe this coincided with the discovery that play was rewarding. The combination of feeling good and playing with others who may have found themselves moving to the sounds was also pleasurable. That sense of well-being in an otherwise challenging external environment could have had strong evolutionary rewards, enhancing the chances of survival. But of course, this story is highly speculative.

Providing a convincing explanation for music's emergence and development requires not only some intelligent guesswork but also depends upon the evolutionary model used, in particular at what point one has to take account of cultural factors, notably at what point culture and biology merged. Music's powerful social and cultural role is clear enough today throughout the world, and there is plenty of evidence, as I shall explore below, that music plays a significant role, along with dance and ritual, in many different societies. Maybe that's all we need to know about music's origins. Leaving the answer at this point would certainly save considerable time and effort spent in making speculations, however interesting and stimulating.

Then if music is essentially cultural in origin, it could still have had evolutionary survival value early on, given the evidence of music making at least 45,000 years ago and likely much earlier than this. It also depends upon what we mean by music, as in a number of non-Western cultures music is integrated with dance and movement and collective performance, so that it may not be accurate to be considering a separate evolutionary explanation for what in the West we consider to be music.

Nonetheless, consideration of the various theories of music's evolution may add to our understanding of music's power to affect us at different levels.

The earliest "adaptationist" theory of music's origin is that of Darwin, who noted that music has the power to arouse a variety of powerful emotions. In order to explain these emotional effects, Darwin used observations from sounds made by animals, particularly male birds who use singing as part of their acoustic display to attract a female mate. This is based upon his theory of sexual selection, the selection of traits that promote the ability to attract a mate and pass on the male's genes to the next generation. He also noted that apes have musical abilities, and that the greatest pitch changes appear to be made by male apes, such as gibbons, when soliciting mates. So, for Darwin, music was used by humans, enabling them to "charm" each other, "before they had acquired the power of expressing their mutual love in articulate language".[9] Music's power to charm tapped into the kind of deep emotions and thoughts from the long-past age prior to language use; first there was song, then there was speech.

Geoffrey Miller developed Darwin's ideas further, relating them to the role of music in contemporary society. Imagining our human ancestors living in groups in the African savanna, he describes music as what "happens when a smart, group-living, anthropoid ape stumbles into the evolutionary wonderland of runaway sexual selection for complex acoustic displays".[10] He describes the "design features" of music with or without dancing which make it a sexually selected adaptation, such as it being spontaneous, involving energy, rhythmic movements, and a certain amount of athleticism—all of which would signal to a potential mate about the health, fitness, and potential fertility of the performer. These features are visible in contemporary society with, for example, the sexual magetism of pop music performers, or their power in attracting followers. Thus, music would be a set of sexually selected aesthetic displays. The advantage of this theory is that, "It makes us recognize that any aspect of music that we find appealing could also have been appealing to our ancestors, and if it was, that appeal would have set up sexual selection pressures in favour of musical productions that fulfilled those preferences."[11]

Whether or not one can make direct links with contemporary society, his theory does have an intuitive appeal, in terms of how musical displays and performances may have been encouraged or enhanced, either though evolutionary pressure with early humans or later through culture. However, the theory is less applicable to the more cooperative aspects of music making, where music brings people together rather than becomes a site for individual sexual display and competition. Brown[12] also makes the point that there is too much about music making that reveals its essential role in group functioning to accept the sexual display theory of music's origins, and that ethnomusicological research suggests that the principal function of music making is to promote group cooperation, coordination, and cohesion. Music also has two distinct design features that reflect an intrinsic role in group cooperation rather than competitiveness—its capacity for pitch blending and metric rhythms. This contrasts with speech, which requires individuality and non-simultaneity to be intelligible.

By being able to blend different pitches into harmonies, music can promote the simultaneity of different parts, thus promoting or encapsulating cooperative group performance and interpersonal harmonisation or group communion. While musical metre is the quintessential device for group coordination, or what was outlined in Chapter Two as entrainment, the combination of pitch blending and metre is highly effective in promoting simultaneous singing and dancing.

Consistent with Brown's views about music aiding group coordination and cooperation, Ellen Dissanayake[13] proposes that such coordination has its evolutionary origins in mother–infant communications, both vocal and physical, the sort that have been described in the previous chapter when considering the notion of communicative musicality. She argues that there is good reason to consider that mother–infant interactions are composed of elements that are literally, not just metaphorically, musical. With their rhythmic and dynamic changes, they are the prototype "for a kind of fundamental emotional narrative that adult music, dance movement, and poetic language can grow out of, build upon, exemplify and sustain".[14] In her thinking, then, the capacity for music evolved not from sexual display but from love or mutuality as seen in communicative musicality.

One could also say simply that the more the mother is attuned to the baby, the more likely a better bond between them and, thus, better chance for the baby's survival. While most caregivers cross-culturally know that singing to their babies is pleasurable or comforting, not all mothers feel they can bond with their babies, particularly if they are depressed or have mental health issues. While most mothers are primarily preoccupied with their newborn babies, this is not an automatic response, just as the wish to have babies is common but not universal. However, babies clearly prefer infant-directed singing to adult-directed singing from very early on,[15] and therefore their response to musically based communication seems innate.

One should add that the particular nature of mother–infant interactions is rooted in evolutionary terms to the changes consequent on humans becoming bipedal. Because of bipedalism, the human female pelvis has a narrow birth canal, with the result that human infants are born relatively early in their development and as a consequence have a long period of helplessness and dependency on their caregivers. Mothers and infants have thus to be able to communicate effectively and quickly long before the baby's speech and language have developed.

Bipedalism also required a larger brain and more complex nervous system to manage the complex sensory and motor actions involved in rhythmic movement needed for walking and running, as well as the complexity of social interactions. The latter was probably the most significant factor in evolving means for vocal communication. Vocal communication was also aided by the changing anatomy of the human larynx, which became bigger and lower in the throat with the change to bipedalism, increasing the diversity of possible sounds it could produce. Along with the change to the position of the larynx, the change to the hyoid bone also became crucial. The hyoid bone helps elevate the larynx and support the tongue whenever a person swallows or talks. Even though the hyoid bone occurs in animals, its unique positioning in humans allows the bone to work in conjunction with the larynx and the tongue to articulate a wide range of distinctive sounds. The presence of an ear canal of modern proportions well over a million years ago suggests that already sounds had become significant for early man.

One can thus see a complex web of anatomical, individual, and social factors facilitating the evolution of human communication, with a likely constant interaction and feedback loop between them all adding to the rapidity of human sociability and connectedness.

Steven Mithen, in his book *The Singing Neanderthals*,[16] advocates the notion of a kind of musical protolanguage, which he calls "Hmmmmm"—standing for holistic, manipulative, multi-modal, musical, and mimetic—which evolved into the separate lines of music and language. He suggests that early humans communicated using Hmmmmm, with gestures, body language, and mime, combined with holistic utterances, many of a highly musical nature. He also has a place for the need for an intense mother–infant communication as a selective pressure for the evolution of Hmmmmm and ultimately for the capacities for music and language in modern humans.[17]

He argues that appreciating that early humans had a sophisticated vocal communication of this kind before music and language helps to explain the archaeological and fossil record. For example, the skeletal remains of Neanderthals suggest a capacity for vocal communication similar to humans, whereas the archaeological evidence provides few traces of what would infer linguistically mediated behaviour, such as a relative absence of symbolic artefacts, though it must be said that they were very adept at tool making.

Mithen attempts to resolve this dilemma by suggesting that Neanderthals did have a complex form of vocal communication but that it was a type of Hmmmmm rather than language, hence his title of "singing" Neanderthals. The separation of Hmmmmm into language and music used by the more cognitively flexible *Homo sapiens* involved their use of the segmentation of holistic utterances. He borrows this notion from Alison Wray,[18] who uses the term "segmentation" to describe the process by which humans break up holistic phrases into separate units, and then combine them into other units in order to produce complex meanings. Such a process would presumably require advanced brain capacity. Also, once such techniques were used to create complex meanings, that in turn would have a knock-on effect on the pace of human evolution and brain development, enhancing cognitive flexibility and fluidity. Mithen postulates that from the original Hmmmmm communication, language developed as a system specialising in the transmission of information, and music developed separately as a system specialising in the expression of emotion.

Mithen's narrative has come in for a lot of criticism for being too general, with the use of a variety of different explanations for musical behaviour, ranging from social bonding to mother–infant bonding and sexual competition. On the other hand, it may well be the case that the most effective way to account for music's evolution is to use a variety of approaches, each of which may carry one important part of the puzzle, in an area where speculation is anyway inevitable. In addition, by emphasising the role of the voice as a crucial element of human evolution, he is pointing to the formative role of basic bodily communication.

Gary Tomlinson[19] provides a magisterial overview of the gradual evolution in parallel of language and music, until their later convergence, through interpretation of data from tool making and settlements over a million years or so. The end products of music and language were preceded by patterns of sociality and communication neither musical nor linguistic, but then coalesced at some point late in human evolution, probably by about 100,000 years ago.

Crucial to this evolutionary process was the increasing organisation of hominin tool making or "taskscaping", the source of increasing cognitive and social complexity, which finally resulted in language and music production as separate but interconnected developments.

According to Tomlinson, the taskscape

> emerges from the varied actions of a social group, the mobile performances of these actions, their structuring of the lived environment, and … the sounds they make. Over against the stable, seen features of the landscape, the taskscape is not external and static but changeable and manufactured; it is not so much seen … as made and heard. The taskscape creates, from the rhythms of action sequences that form it, its own temporality, one based on moments of mutual attention

commanded among its participants by movement and gesture. It describes hominin movement through time connecting the material and the social.[20]

To make his case, Tomlinson covers in great detail evidence of early human tool making, from over one million years ago, which even then required significant skill. At that time, repetitive action sequences to make stone tools reflected already the "rhythmitization of brains, bodies and hands inscribing the world and giving dimensions of space and time to daily life".[21] There was no music or language even by 500,000 years ago, but only a set of action sequences, partly vocal, under greater control and of greater communicative utility than before.

There was a long period of stability with regard to tool making until about 300,000 years ago, when more complex tool making gradually began to appear; this seemed to coincide with a fundamental change in social behaviour, when hominins began passing on information and skills within the group, rather than having to basically work things out for themselves ad hoc, on a day-to-day basis. This change in itself appears to have acted as a positive feedback loop, greatly increasing hominin evolution. The emergence of complex human capacities entails not only biological changes but also involves shifts in social behaviour which in turn impacts on the rate of evolutionary change. In order to transmit information these early hominins may well have been mirroring their complementary gestures, and already were beginning to show features of social entrainment, reflecting changes in neuronal connectedness, such as increasing coordination of motor patterns, prefiguring what would in time emerge as musical rhythmic cognition.

The environment was likely by then to be a "voicescape", given hominin changes in auditory and laryngeal anatomy, though not yet involving language.

The idea of a feedback loop between human cultural invention and biological evolution makes Tomlinson's a biocultural theory, meaning that already early on in human evolution, culture had a significant part to play in facilitating change.

Thus, Tomlinson discerns links between the social and technological lives of hominin species that preceded us, based upon the archaeological and palaeontological records, and musical capacities that took shape much later. He assumes that the capacities for music processing are innate in the human brain, probably involving a number of different domains and neural networks, even though learning and acculturation play a large role in any musicking. Rather than seeking a unitary explanation for music's evolution along the lines of the previous theories I have covered, he aims for an incremental theory, where musicking involves many cognitive and cultural changes, which indeed point to a long, varied, and piecemeal development.

He suggests that evolution essentially "exploded" in the last 100,000 years, at which point there occurred the final coalescing of species-wide capacities for musicking, language, and other human activities.

He sees musicking as essentially always technological. "Its modes of cognition were shaped from the first by the extensions of the body that were the earliest tools and weapons, in ways that left a deep imprint on both sociality and the genome. Musical instruments as such came later, but this broader, crucial instrumentality appeared long before there was music."[22]

But music was also always social and the mix of the technological and the social formed the matrix in which musicking took shape. Tomlinson points to a complex interaction between music and language evolution; it is not that easy to disentangle them, with there being much in common as well as some differences. However, it is clear that the discrete pitch processing of music does not occur in language. Musicking was more linked to emotional communication, but went hand in hand with an increasing development of the theory of mind, with the recognition of other humans. Gesture calls, not unlike Mithen's Hmmmmm, probably preceded language and music and may have involved increasing mimetic entrainment, turn taking, and sharing of attention. In any case, vocalisation preceded musicking and language. "*Musicking in the world today is the extended, spectacularly formalized, and complexly perceived systematization of ancient, indexical gesture-calls.*"[23] Musicking coalesced from "a range of cognitive and perceptual capacities alongside a coalescing social complexity".[24] This complexity increasingly involved thinking at a distance, the notion of other minds, and the elaboration of imagination, the ability to imagine other worlds.

My reading of this account is that the gradual building up of a home base through a process of what one could call "*homescaping*" became one crucial element of this evolution. One can see in the data how tool making became slowly but increasingly sophisticated as loose gatherings of hominins began to form local niches then more organised settlements, where, as I mentioned, skills could be passed on rather than created ad hoc.

Music may have evolved as a powerful means of enhancing intersubjective communication as early humans began to gather into loose and then more organised communities and simple homes. Music could then be seen early on as a unifying influence. I suspect it would be difficult to determine quite when the *maternal voice* came to have such a crucial role in organising the baby's experiences; this issue is not covered by Tomlinson, who is more focused on social communication in general.

This notion of homescaping in part accounts for how music reaches powerfully into the depths of our psychic rootedness, into the interior of our souls or the "other world" of our unconscious. Music can thus be seen to be vitally linked to the very ground of our being as a direct result of its evolutionary origins; and it shows at an early stage in our evolution the subtle connections between music, home, and community that one can see in many cultures today.

Language and music, though interconnected both in their origin from some sort of early gesture calls, increasing social connectedness, and biological and cognitive complexity, also took on different roles.

> In summary form, the systematized abstraction of the symbol came to inhabit language, while musicking came to be governed within a loose assemblage of formalized performative abstractions and systematized indexes foreign to normal linguistic use. These differences are fundamental, and they shape to this day what societies everywhere recognize, in tacit practice or explicit formulation, as the special powers of musicking. Linguists may rightly tell us of the sovereign mysteries of language structure and process; but only *lift a voice in song*, and all humans are struck—enthralled, seduced, threatened, made, or unmade—by these powers ... [H]owever these differences appear in our developing understanding of them, they will be best understood from the context of a deep-historical inter-relation and coevolution.[25]

Ian Cross similarly sees music and language as "subcomponents of the human communicative toolkit—as two complementary mechanisms for the achievement of productivity in human interaction though working over different timescales and in different ways".[26]

Given that music has what Tomlinson describes as special powers, for example, to engage human emotion, one can ask what evidence there is for the universality of such powers across different societies and culture, and what this tells us about the nature of these powers, and how this may link with the various uses and functions music plays in human society. My understanding of this complex issue is that there are indeed certain universal elements in music making, as there are indeed with language use, and this is supported by cognitive research and neuroscientific findings, but there are also many variations in how music is performed and in its social and cultural use, as shown by ethnomusical research. It would be thus incorrect to see music as merely a product of human culture, or as a mere side effect of language, or to ignore the biological substrate on which musical processing and understanding are based, while it would also be short-sighted to ignore the incredible richness in how music, with or without dance and words, is performed across the world.

Music's origin in biology

There is considerable evidence for the biological foundation of music perception and processing, not just from the evolutionary evidence just covered. First of all, there is the mother–infant work already described, where there are many examples of infants' early musicality, where biology interacts with early intersubjectivity. As I mentioned, babies learn and tune into the characteristic prosody (rhythm and melody) of their own language, but they can also discriminate in melodies, pitch intervals, and the timing of pitches. As Sandra Trehub has described, infants, like adults, when presented with a novel melody in different ways, focus mainly on the pitch contour, the ups and downs or pattern of intervals, as well as the rhythm, reflecting a basic disposition to attend to relational pitch and timing cues rather than to specific pitches and durations.[27] Infants and adults also retain more information from sequences whose component tones are related by small-integer ratios rather than large-integer ratios. Infants are also able to recognise transposed melodies, that is, when they are shifted up or down in pitch, and this capacity develops spontaneously by six months.

Early studies revealed that infants preferred consonant to dissonant intervals, but more recent studies with children from ethnically and culturally diverse families did not do so, as they probably heard a greater variety of music. This indicates that even if infants are born predisposed to prefer auditory consonance, this can be modified by experience.[28]

Across cultures, infants respond particularly positively to infant directed speech and to lullabies; singing is more effective than speech in calming infants.[29] The vocal play of six to twelve month olds that leads to singing is clearly distinguishable from the vocal play associated with pre-speech, both in its use of stable pitch levels on vowels and its rhythmic organisation.

Thus, overall, one can assert with some confidence that normal babies are born musical and that they display remarkable musical abilities and responsiveness to music that in some ways are similar to that of adults. There is good evidence that human infants are well equipped to learn the musical regularities of their environment; they seem to have a "music-detector" system,[30] well attuned to a potential caregiver.

Further evidence for the biological disposition for musicality and its autonomy comes from insights gained from those children and adults who are congenitally unable to process much musical information—from congenital amusia,[31] as described, for example, in the research of Isabelle Peretz. Amusia affects about 2.5% of the population, and can be diagnosed by a test which requires participants to discriminate pairs of melodies that may differ by a single tone that is out-of-key. There are also about 3.5% of the population who cannot detect an off-beat tone.

This disorder affects music but not speech. Amusia runs in family and so has a strong genetic component. Individuals affected have mainly normal receptive and expressive speech and language. But they have difficulty in recognising instrumental melodies, or hearing when someone sings out of tune or plays a wrong note. They can recall the lyrics of a song but not the tune, so that there is a clear-cut dissociation between music and speech. In addition, however, they also have some difficulty in detecting large pitch movements even in speech, so that they may find difficulty in detecting if a statement is a question and when the ups and downs of the pitch movements are larger rather than smaller. Thus, the deficit is in the tonal encoding of pitch, which obviously would generally affect musical more than speech perception. Evidence from brain imaging research reveals problems in the connectivity between parts of the auditory cortex and the frontal cortex—a problem with the right fronto-temporal pathway, or there may be dysfunction in the pre-frontal cortex.

Further evidence for the autonomy of musicality comes from the effects of brain damage in adults, where brain lesions can selectively interfere with or affect musical abilities in a variety of ways. Some brain lesions, such as a stroke to the right hemisphere, may lead to acquired amusia, with similar problems to those with congenital amusia. The existence of a "specific problem with music alongside normal functioning of other auditory abilities, including speech comprehension, is consistent with damage to processing components that are both essential to the normal process of music recognition and specific to the musical domain".[32]

Aphasic patients with damage to the left hemisphere may be able to sing familiar tunes while being unable to produce intelligible lyrics or even speak. There is the famous case of the Russian composer Vissarion Shebalin, described by A. R. Luria, who had a left temporal and parietal stroke leading to severe aphasia, but carried on teaching and composing music of quality.[33]

As Oliver Sacks described in vivid detail, different brain lesions may lead to various distortions, excesses, and breakdowns in musicality.

> The power to perceive (or imagine) music may be impaired with some brain lesions; there are many forms of amusia. On the other hand, musical imagery may become excessive and uncontrollable, leading to incessant repetition of catchy tunes, or even musical hallucinations. In some people, music can provoke seizures. There are special neurological hazards, "disorders of skill", that may affect professional musicians. The normal association of intellect and emotion may break down in some circumstances, so that one may perceive music accurately, but remain indifferent and unmoved by it, or conversely, be passionately moved, despite being unable to make any "sense" of what one is hearing.[34]

Sacks also highlights a particularly interesting congenital condition, which gives some further evidence of the existence of a specific brain capacity for music—Williams Syndrome, caused by a microdeletion of 26–28 genes on chromosome 7. Children with this syndrome may have a variety of issues, including cardiac defects, social communication difficulties with delayed development, and some intellectual impairment as well as being typically overly friendly. In addition, Sacks describes their frequent heightened musicality, despite other deficits, with an unusually high degree of engagement with music. Even when such a child has developmental delay they can still be adept at picking up tunes or matching pitches with a parent on the piano. There appear to be specific brain differences compared to normal, including an enlarged neocerebellum. The cerebellum is concerned with coordinating movement, and may be thus involved with musicality, with the coordination of pitches and rhythms.

Though I shall cover further neuroscientific findings relevant to music processing and also the link between music and emotion in the next chapter, one can say in general that the brain seems hard-wired for music, and this is separate from the capacity for language, though there is likely to be significant overlap is some key areas such as responses to pitch and timing. Both music and language may well share some mechanisms, what Aniruddh Patel calls "resource sharing".[35] And music processing seems to recruit "… a vast network of regions located in both the left and right hemispheres of the brain, with an overall right-sided asymmetry for pitch-based processing".[36]

Music and language share some common structures, such as the grouping of musical or linguistic events into coherent chunks, notes, or phrases; stress patterns, metre, contour, and timbre, or tone colour. Music differs from language in being non-referential, lacking a semantics, organised in terms of blendable pitch variations and tonal tensions, while language has syntax and meanings, and phonologically distinctive features. This is further suggestive evidence that such differences are reflected in different brain pathways, once more an indication that there is a significant biological contribution to the foundation of human musicality. But, as John Blacking describes, based upon his study of African music, musicality also has to be seen in a cultural context if we are to understand its emotional power. "[I]f we take a world view of music, and if we consider social situations in musical traditions that have no notation, it is clear that the creation and performance of most music is generated first and foremost by the human capacity to discover patterns of sound and to identify them on subsequent occasions. Without biological processes of aural perception, and without cultural agreement among at least some human beings on what is perceived, there can be neither music nor musical communication".[37]

Music's origin in culture

While music has the power to awaken in our bodies all kinds of responses, people's responses to music cannot fully be explained without some reference to their experiences in the culture in which they are at home. By studying how music is understood and performed in different societies, one may find commonalities.

Virtually all human societies have various forms of musical activity, though not all name this activity as "music", as it may also include in one concept dance, ritual, and song. Most native American languages and several African languages have names for individual music genres, such as singing, playing instruments, and dance, but not an overall name for all genres. The name for music then often includes other activities with no clear distinction between them, while each society or tradition has its own particular conception of music.[38]

Music cannot be defined just by sound alone, as it involves human beings in social interaction, made by people for other people.[39] There are however certain "statistical" universals,[40] such as innate musicality, the existence of tradition-carrying networks, the ubiquity of distinct repertoires of children's music,[41] and the presence of musical units or phrases, identifiable by, for example, some kind of repetition, the latter being present across musical systems throughout the world. Reports from cross-cultural studies show that listeners from different parts of the world can relate emotionally to unfamiliar music from different cultures, suggesting the universality of musical emotions.[42]

Musical phrases, or combinations of phrases, frequently make use of short intervals such as seconds or thirds, as well as the pentatonic scale; while most musical systems do not use more than seven pitches in an octave. All known cultures accompany religious activity with music,[43] and music is very commonly associated with the supernatural or "other worlds". All cultures use music to accompany dance, and all songs have words. Certain ideas about music are frequently present across the world.[44] These include a notion of a culture's music being natural or directly linked with nature, or linked to the cosmos or universe in some way. Music is believed everywhere to affect our emotions and often felt to be transformative. I have already covered the role of early communicative musicality, which appears to be a universal phenomenon, albeit with some cultural variations in its expression.

Alan Merriam, based on his review of world cultures, lists ten essential functions of music within any society—to express emotions, for aesthetic enjoyment, for entertainment, for symbolic representation, to encourage physical reactions such as dancing, war making, or hunting, for enforcing conformity to social norms, as a validation of social institutions and religious rituals, to contribute to the continuity and stability of a culture, and to facilitate integration of society.[45] As can be seen these functions do overlap, with the main focus being on social and cultural cooperation at a group level.

> Music … provides a rallying point around which the members of society gather to engage in activities which require the cooperation and coordination of the group. Not all music is thus performed … but every society has occasions signalled by music which draws its members together and reminds them of their unity … Music is clearly indispensable to the proper promulgation of the activities that constitute a society; it is a universal human behaviour—without it, it is questionable that man could truly be called man, with all that implies.[46]

These considerations imply that in order to understand the role of music in different cultures, one needs to pay attention to *group processes*. Much music making occurs within what Freud, in *Group Psychology and the Analysis of the Ego*, called "short-lived" groups, as opposed to long-lived groups such as an army or a religion;[47] though a number of short-lived groups which come together to make music do so as part of a religious or cultural event such as mourning for a dead member of society or to mark the passage from one state to another such as in initiation ceremonies. It is natural in groups for there to be a reduction in conscious rational thinking, a tendency towards the predominance of primary process activity, and an increase in the intensification of affects. That certainly seems to be the case with those involved in trance states facilitated by musical activity. However, as I shall describe below, it is noticeable that the musicians who create the music usually remain outside the trance experience, as they need to keep intact their ego functioning on behalf of the rest of the group.

Music-making groups appear to be similar to those which Freud described as "primary" groups, that is, those groups "that have a leader and have not been able by too much 'organization' to acquire secondarily the characteristics of an individual".[48] Freud wrote that the main forces keeping the group together were particular kinds of libidinal ties, where the sexual drives are aim-inhibited, that is, through identifications; the latter are with each other and with the leader—though, aside from Western concert music, music making often involves some kind of joint leadership, thus diluting the potentially coercive role which a single group leader can fall into. Shamanism, however, exploits the power of the shaman as leader of the musical activity which accompanies his functions.[49] Music itself, through entrainment, has a powerful role in sustaining the continuity of the group.

The point is that during the various musical activities in the many different cultures described by ethnomusicologists, there are complex group dynamics, with various kinds of identifications, both with group members and with those outside the musicking group.

Émile Durkheim, in his book *Elementary Forms of Religious Life*,[50] had already shown how intermittent states of stimulation occur when groups of people congregate together in crowds and assemblies of various kinds. Basing himself on the religious life of so-called primitive societies, he described how the act of congregating together

> is an exceptionally powerful stimulant. Once the individuals are gathered together, a sort of electricity is generated from their closeness and quickly launches them to an extraordinary height of exaltation. Every emotion expressed resonates without interference in consciousnesses that are wide open to external impressions, each one echoing the others.[51]

But these intense social feelings would be quite unstable without the existence of symbols that provide a structure such as an emblem on which they can fix. I would add that it is not only symbols that provide such a supporting structure for such feelings but also structured events such as rituals, rites, and beliefs.

Such complex group dynamics and ways that music interacts with the group can be seen in the triggering, sustaining, and closing of *trance* states.

In his classic account of music and trance, Gilbert Rouget points out that trance is a phenomenon observed throughout the world and is associated most of the time with music; the aim of his book is to answer why this is the case, focusing mainly on religious trance.[52] His finding is that music is the principal means of manipulating the trance state, by "socializing" more than triggering it. The process of socialisation varies from society to society, and takes place in different ways, depending upon the particular society's belief system; in each case a different logic determines the close relationship between the music and the trance state.

Trance is to be distinguished from ecstasy, which takes place generally without music, in silence and solitude, with sensory deprivation and in a state of immobility. Trance in contrast requires movement, noise, occurs in company, with sensory overstimulation, and, unlike ecstasy, involves amnesia.[53]

The manifestations of trance can resemble classic hysteria, or even hypnosis, with trembling, shuddering, foaming at the mouth, protruding eyes, convulsions, falling to the ground, analgesia, and so on.

There are two main types of trance—possessive and shamanic. In the former, the inhabitants of the invisible "other world"—gods or spirits, visit the participants; in the latter, the shaman makes a journey to the spirit world in a kind of séance or "psychic journey". The purpose of both forms of trance varies with the culture, but with possession there is a wish to meet the spirits as part of a religious experience; with shamanism it may be so that the shaman can heal or bring back news from the spirit world for the good of the community. Sometimes drugs are also used as part of the process of possession. Trance states can enable the adepts to feel in touch with new knowledge or a world beyond the everyday.

There is a certain predictability about trancing behaviour, which probably adds to the ritualistic aspect of the trance. As Judith Becker, in her book on trance describes:

> Trancers behave exactly the way in which they have learned to behave, and trance behaviour is narrowly circumscribed by time and place. In my hometown, Pentecostal trancing can be witnessed each Sunday morning sometime between 11.30 a.m. and 1.30 p.m., in a particular church, with all trancers following basically the same gestural script. Likewise, in the [Balinese] Rangda/Barong ceremony the trancers who attack Rangda [a masked witch figure] and then turn their knives against themselves do so on cue ... Very rarely do they enter trance before the proper narrative moment ... Trancers follow a script that determines the time of the onset of the trance, the duration of the trance, behaviour during the trance, and the style of withdrawal from the trance.[54]

With possession, there are successive phases—preparation, onset, climax, and resolution. Music can play a part in all these phases, and in which phase it does so depends upon the culture, the particular stage reached by the adept—novice or experienced—and by the particular ritual being observed. For example, with regard to the coming out of a trance,

> In Bali, the trance of the little girl dancers is ended through specific songs ... In Chad, among the Mundang, in order to return the possessed person to their normal state, the musicians play a particular

theme while the chief officiant taps them on the back. In the [Brazilian] *candomblé*, it is a rhythmic theme played by the drums that tells the dancers … to leave the room at the end of the ceremony.[55]

Generally, the music involved with trance consists of both vocal and instrumental forms, though the relative importance of one or the other varies from one cult to another. The instruments are usually either melodic, such as the fiddle, or rhythmic like the drum. A musical theme or themes may have to be played in order to trigger the trance; it is not triggered by musical rhythm itself but needs the contribution of beliefs and imagination, such as calling on the spirits. The ceremony consists of the "musicants", the adepts and the spectators who help the adepts attain a trance state, and the musicians, who may well be professionals hired for the occasions, and who do not lose their sense of self. The music cannot be performed by the adepts themselves as they have to be "subject to" the possessive process in order to go into the special trance state of mind, somewhat similar to the hypnotic state, though not identical with it. The latter does not require music.

Rouget describes one of the main functions of music in these situations is to "establish communication with the gods and thus to create a situation of identification and alliance that favours the kindling of possession".[56] Thus, music sustains a powerful group dynamic. It works, not just because of its physical effect, which by itself will not produce a trance, but by means of its "moral effect", because the music is in the service of a belief system. However, music is particularly effective as a corporeal means, with or without dance, of facilitating the trance state. As I shall discuss in the next chapter, the likely physical reason for this is that it directly involves the basic instinctual–emotional core of the brain, the brainstem and midbrain, responsible for primordial feelings. But trance states in addition require "a certain conjunction of emotion and imagination. This is the source from which trance springs."[57]

Judith Becker further tackles the nature of trancing, which she sees as a process simultaneously physical and psychological, somatic and cognitive, and she involves neuroscientific explanations in her comprehensive study of the field, adding to the basic findings of Rouget.[58]

Becker describes how

> Trancing can be empowering for all concerned, attesting to the divine presence in one's midst, legitimizing the religious beliefs and

practices of the community, and often bestowing deep satisfaction on the individual trancer. Trancers are always active and often dance—a response in part because of their deep emotionality. They welcome emotion, they offer themselves to emotion as they enact emotion. Musical immersion stimulates emotion and facilitates their special attentiveness, their special consciousness.[59]

It seems to open up pathways not usually experienced in daily life.[60]

Linked closely to the deep emotionality of the trancers, and indeed those witnessing or facilitating the trancers, for the trancing is part of a whole group or community experience, is what Becker calls "*deep listening*".[61] This is a descriptive term, borrowed from the composer Pauline Oliveros, when a listener pays close attention to what is going on below the surface of what is heard. This seems similar to close psychoanalytic listening, where one is responsive to unconscious or preconscious communications. For Becker, it refers to those who are deeply moved, even to tears, by simply listening to music. She calls it a form of secular trancing,

> divorced from religious practice but often carrying religious sentiments such as feelings of transcendence or a sense of communion with a power beyond oneself. Deep listening may be attributed to personal psychology as in the United States, or may be culturally situated. So closely are music and deep listening associated in Arabic cultures that the term *tarab*—which means to be moved, agitated, while listening to music (to the extent that one may cry, faint or tear one's clothes)—can also simply refer to a musical style ... Like *tarab*, "deep listening" conjoins musical expression and the emotional impact of musical expression.[62]

I would add that deep listening implies a particular form of attention, rather like the analyst's free-floating attention, or what Anton Ehrenzweig called "unconscious scanning",[63] paying attention to the undifferentiated depths of the mind, where subject and object become merged and undifferentiated. Trancers seem to be able to reach these inner depths, thanks to the support of music. They reach states where their sense of self appears lost or merged with the group, but are then in touch with what they experience as a transcendent world, that of God, gods, or spirits. The secular equivalent could be that feeling of spirituality and transcendence reached when listening to any deeply felt music. Thus, imagine hearing the Vespers of Monteverdi in a beautiful church as sung by The Sixteen under Harry Christophers. One feels heightened emotion and tension, the enjoyment of the sounds, harmonies, and tonal relationships, with their consonances, dissonances, expectations, and surprises, all the excitement of the musical shifts and modulations. In a sacred or special place with a great conductor there is an added element, a new depth and heightened understanding of some essential aspects of human life. For those with religious belief there is the added effect of the meaning of the words. As in any dramatic performance there is a concentrated vision of human life through the artistic medium, affecting us at a number of levels, consciously and unconsciously, penetrating the soul. Though not a trance state, in that such listening retains the ordinary conscious controls, it is nonetheless not that distant in terms of the musical experience and the state of heightened arousal attained in the act of deep listening.

Of course, to allow oneself to go into a religious trance state, where the sense of self becomes merged, requires a particular context, what Becker describes as the "*habitus*" of listening. This comes from the thought of Pierre Bourdieu, for whom "habitus" refers to a person or group's durable transportable dispositions, a tendency to behave in a certain way, their ways of being and their feel for their own field of cultural practice.[64] This notion is chosen to account for how modes of listening "vary according to the kind of music being played, the expectations of the musical situation, and the kind of subjectivity that a particular culture has fostered in relation to musical events".[65] That is, our perceptions of musical emotion take place within a set of habits and modes of being related to our own cultural practices, and, I would add, psychic home, the ground of our being and subjectivity.

To illustrate this notion, Becker mentions the similarity of the habitus of listening of the listener to Hindustani music and the Western listener in terms of physical stillness, focused attention, and inner withdrawal, and that both musical traditions claim to represent emotion.[66] This compares with, for example, the habitus of listening of the Wolof griots of Senegal, where the musical expression of emotion is dialogical and situational, not personal and interior, while the griots are highly expressive and excitable while musicking.

The trance state also requires a particular model of subjectivity, which allows the self to lose body boundaries. The Western notion of the human subject, with the sense of a bounded, unique, and self-contained personal centre, may well hinder the ability to surrender to the trance experience.[67]

Becker brings in various neuroscientific findings to help account for the trance experience. Though I shall leave tackling the details of such findings to the next chapter, it is worth mentioning at this point the important role of physical and emotional musical entrainment as facilitating powerful group emotion, coupling the mind and body of the trancer to that of those involved in the group ritual and to the overall habitus of the proceedings.[68] This coupling process can be accounted for by the way that networks of neuronal pathways can be connected and reconnected at the level of the brainstem and midbrain, responsible for much emotional production and regulation.

Steven Feld's studies concerning music making among the Kaluli people of Bosavi in Papua New Guinea adds further evidence concerning the intimate relationship between music and emotion, and how musical structures can channel emotion. He describes how "By analyzing the form and performance of weeping, poetics, and song in relation to their origin myth and the bird world they metaphorize, Kaluli sound expressions are revealed as embodiments of deeply felt sentiments".[69]

The main form of the bird myth (one of several) that structures the study is "The Boy who became a Muni Bird". This concerns a boy and his older sister, who go off to catch crayfish. The sister is successful while the boy is not, but asks repeatedly for his sister to give him a crayfish, which she declines—this goes against the Kaluli value of care for others. In the end, the boy catches a tiny shrimp and turns into a muni bird, one of the fruitdove family of birds surrounding the Kaluli. As the sister becomes upset and asks the bird not to fly away, the bird/boy sings its high falsetto cooing cry and flies off.[70]

The myth provides a statement about birds, social values, and soundmaking, all of which are central to Kaluli life. The various birds in the rainforest are classified by the Kaluli depending on both their physical features and the sounds that they make. The muni bird is central in their musical life, with a three- or four-pitch descending melodic contour, which comes to stand for loss, such as death, or the sound of a child abandoned, hungry, and alone. In the myth recounted there are two types of weeping when the boy turns into the bird—a fast, hysterical crying and a more definite pitched, slower melodic crying. The social organisation for weeping in Bosavi is similar in its melodic contour. The three- or four-note melody is used as a sound metaphor for sadness, expressing the sorrow of loss and abandonment.[71]

In the Kaluli belief system, birds are "voices" "because Kaluli recognize and acknowledge their existence primarily through sound, and because they are the spirit reflections ... of deceased men and women".[72] They represent the spirits of the dead, as in the boy/bird myth. For the Kaluli, sound is a dominant cultural means for making sense of their world; they are surrounded by the rich sound world of the rainforest. Bird sounds metaphorise Kaluli feelings and sentiments. Indeed, when Kaluli perform weeping or song, they so strongly identify with the birds that they become birds, and when others evaluate how moving their performance has been, they compare the performers with the birds. Bird sound words are invoked to express a variety of complex emotions involved with loss and abandonment.[73]

Thus, "becoming a bird" is the core Kaluli aesthetic metaphor because it embodies the emotional state that has the unique power to evoke deep feelings and sentiments of nostalgia, loss, and abandonment.[74] Organised and synchronous sounds as songs, or songs with accompanying instrumental music, are an integral part of that emotional state.

Here once more is clear evidence that a principal function of music making is to promote group cooperation, coordination, and cohesion through the evocation and channelling of emotions, some of which at least are both specific and complex emotions involved with complex states of mind.

CHAPTER FOUR

Music and emotion, first movement

I have touched upon many descriptions of music's direct connection with emotion. In this chapter, I shall look in more detail at that link, focusing on the nature of emotions from a variety of different perspectives. I believe that with such a complex field as that of human emotion, there is a danger, driven by the desire to try to make sense of such complexity, of being seduced by one kind of explanation. At the moment, neuroscience has become fashionable. When information from a brain scan is shown, it seems to provide a "real" explanation for human behaviour, or an explanation that is not open to question, as if other hypotheses from, for example, human encounters and social and cultural life are of doubtful significance. In my view, we need to include a variety of different ways of approaching the study of human emotions if we are to find a way through the web of complex and at times apparently quite mysterious ways that music affects us; that also means dropping the Cartesian view of the split between mind and body and accepting understanding from both the human and the physical sciences, with each making a significant contribution to our overall understanding of music's power to affect us.

I shall begin with my own discipline of psychoanalysis, not only because it is one with which I am familiar, but also because in the psychoanalytic encounter we are often faced upfront with the dilemma of human emotion at its most taxing. In this, we constantly come across ambivalent, cut-off, or denied feelings, or various mixed states, and from these, there is a great deal to learn about the nature of affective life, and hence how music may in turn influence our states of mind. The psychoanalytic theory of affects, particularly as put forward by Freud in his early explorations of mental illness, also link up with current neuroscientific findings; perhaps not surprising, given that he started out as a neurophysiologist.

Psychoanalysis and emotions

In psychoanalytic practice, we are constantly dealing with how the patients' emotional life impacts on their relationships. Analysts have to deal with the complexity or messiness of their patients' emotions, without the benefit of controlled studies or a discrete focus on only one aspect of their emotions and feelings. The psychoanalytic approach to the study of emotions, within daily practice of analysis, reveals an incredibly complex situation, where emotions, feelings, and ideas have to be taken account of in their various interactions. Examination of the psychoanalytic understanding of emotions is potentially an important basis for appreciating the way that music links up with emotions and the body at various levels, and, as I shall argue, links up reasonably well with *some* recent neuroscientific findings, though the latter often fail to take account of the presence of ambivalent emotions and conflicts between emotions, or the fact that emotions can be disturbing, disrupting and fragmenting, overwhelming or deadening—phenomena with which the psychoanalyst in the consulting room is only too familiar.

It is worth noting that there is considerable variation between and within disciplines in the use of the basic terminology concerning emotional life, and each set of terms is related to the particular model of the mind being used. For that reason, it is probably wisest to have a flexible attitude to such definitions, as there is anyway considerable overlap between terms. For the sake of an initial clarification, one could see the terms "*affect*", or "*emotional experience*" as the most general terms for the whole field, incorporating bodily sensations and more complex subjective experiences. Affect is sometimes used specifically to describe the experiences of desire, but is usually a general term for any emotional state or mood. Feelings often refer to any experience of the self or subject accompanied by bodily sensations. For the philosopher Susanne Langer, "Feeling, in the broad sense of whatever is felt in any way, as sensory stimulus or inward tension, pain, emotion or intent, is the mark of mentality."[1] She saw the entire psychological field as a vast and branching development of feeling".[2] She also thought that "The real power of music lies in the fact that it can be 'true' to the life of feeling in a way that language cannot: for its significant forms have that *ambivalence* of content which words cannot have."[3]

There may be *short-lasting* emotional experiences, which could be described as *feelings* or *emotions* or even moments of arousal, and more *long-lasting* emotional experiences which are generally described as emotions rather than feelings, and often involve a complex set of relation-ships and experienced events—what I would call an "emotional journey". People can talk about acting or performing with "feeling", or being "in touch" with feelings, without necessarily referring to a particular state of mind, but more like an attitude where one puts one's "heart and soul" into what one is doing or experiencing. The term *"moods"* tends to be used to describe the general tendency to have an emotional experience, some kind of background "tonality", or basic feeling state, and can persist for a length of time, but so can an emotion, such as love or anger, which reveals that emotional experiences are often *processes* rather than just static states of mind. There may be some basic emotions, such as hate, anger, disgust, joy, which often involve some kind of typical facial expression, but the more one looks at the context for having such an emotion, the more one comes up against the complexity of human relationships, where emotions are not just basic but involve various amounts of judgement and evaluation, sometimes, but not always, leading to action. Given such a confusing field of terms, I will often just refer to "emotions" or, in general "affects", as the most readily understandable terms for the phenomena under consideration.

From the early days of Freud's clinical investigations, the patients' emotions, or in more general terms what he called their *"distressing affects"*, were crucial in understanding the nature of their symptoms, what brought them to seek help. The concept of affects played an important part of Freud's theory of the psyche from the beginning, with an emphasis on the psychic conflict between ideas having incompatible amounts of affect as lying behind symptoms.

In Freud and Breuer's "Preliminary Communication" about hysteria in 1893, and in *Studies in Hysteria* in 1895, it was clear that in the mind there are *separate paths* for language, or representations, and affects. This finding matches to some extent what I have outlined in the previous chapter, when, for example, Gary Tomlinson has conjectured about the separate evolutionary paths for music and language; or when one takes account of the effects of brain damage, such as when the normal association of intellect and emotion may break down in some circumstances, so that one may perceive music accurately, but remain indifferent and unmoved by it, or conversely, be passionately moved, despite being unable to make any "sense" of what one is hearing.

Freud and Breuer hypothesised that traumatic memories, or severely distressing affects, were at the root of hysterical symptoms. Such hysterical conditions included various bodily symptoms such as persistent neuralgias and anaesthesias of very varied kinds that did not follow real anatomical boundaries, contractures and paralyses, hysterical attacks and epileptoid convulsions, which appeared to be truly epileptic, chronic vomiting and what we would now call anorexia nervosa, and various forms of disturbance of vision, such as constantly recurrent visual hallucinations. They asserted that the origin of the hysterical symptoms was to be found in one or more traumatic events which have been met with a corresponding and proportionate discharge of distressing affect.

They described that their "… investigations reveal, for many, if not for most, hysterical symptoms, precipitating causes which can only be described as psychical traumas. Any experience which calls up distressing affects—such as those of fright, anxiety, shame or physical pain—may operate as a trauma of this kind."[4]

Freud and Breuer presumed that the

> … psychical trauma—or more precisely the memory of the trauma—acts like a foreign body which long after its entry must continue to be regarded as an agent that is still at work; and we find the evidence for this in a highly remarkable phenomenon which at the same time lends an important *practical* interest to our findings. For we found, to our great surprise at first, that *each individual hysterical symptom immediately and permanently disappeared when we had succeeded in bringing clearly to light the memory of the event by which it was provoked and in arousing its accompanying affect, and when the patient had described that event in the greatest possible detail and had put the affect into words.* Recollection without affect almost invariably produces no result.[5]

Thus, the recall of the traumatic memory brings about the revival of the affect which was originally attached to it, which leads to the removal of the hysterical symptom; this is the "talking therapy".

They made the point that it was puzzling why events experienced so long ago should continue to operate so intensely, that their recollection should not be liable to the wearing away process to which memories normally succumb. This fading of memory, or the loss of the affect, is often caused by a particularly

> ... *energetic reaction to the event that provokes an affect.* By 'reaction' we here understand the whole class of voluntary and involuntary reflexes—from tears to acts of revenge—in which, as experience shows us, the affects are discharged. If this reaction takes place to a sufficient amount a large part of the affect disappears as a result ... If the reaction is suppressed, the affect remains attached to the memory. An injury that has been repaid, even if only in words, is recollected quite differently from one that has had to be accepted ... language serves as a substitute for action; by its help, an affect can be 'abreacted' almost as effectively. In other cases, speaking is itself the adequate reflex, when, for instance, it is a lamentation or giving utterance to a tormenting secret, e.g. a confession. If there is no such reaction, whether in deeds or words, or in the mildest cases in tears, any recollection of the event retains its affective tone to begin with.[6]

One might speculate that part of the power of singing is that affects are literally given a voice; for those who are traumatised, this release of affect and its accompanying memories may be liberating.

Freud and Breuer found that the memories which have become the determinants of hysterical phenomena persist for a long time with astonishing freshness and with the whole of their affective colouring, while they are also, on the other hand, unlike other memories of their past life, not at the patients' conscious disposal, or are only present in fragments. The strength of the affect which can be released by a memory is very variable, according to the amount to which it has been exposed to "wearing away" by different influences, and especially according to the degree to which the original affect has been "abreacted". If the original affect was not properly discharged in a normal, but in an "abnormal" way, the excitation arising from the affective idea is "converted" into a somatic phenomenon, the so-called "conversion hysteria", with symptoms such as hysterical paralysis.

Traumatic memories (seen by Freud, early on in his career at least, frequently of directly sexual origin), with their "affective colouring" or "accompanying affects", were what Freud would soon afterwards describe as being unconscious. They remained unconscious because of the way that the disturbing affect remains *attached* to the memory of a traumatic event. They were however able to reach consciousness by "... *allowing its strangulated affect to find a way out through speech*".[7]

Affects are not necessarily tightly bound to memories, or representations, or ideas, but can go their separate ways, depending upon the kind of psychic condition of the patient. Indeed, Freud listed early on three possible lines of development of the affect: transformation of affect in conversion hysteria, displacement of affect in obsessional neurosis, and exchange of affect in anxiety neurosis and melancholia.

Freud further explores the nature of affects in *The Interpretation of Dreams*.[8] He points out that in dreams the ideational content is often not accompanied by an appropriate affect. "In a dream I may be in a horrible, dangerous and disgusting situation without feeling any fear or repulsion; while another time, on the contrary, I may be terrified at something harmless and delighted at something childish."[9] This form of affect displacement can be seen in a variety of daytime situations of high emotion. A classic example can be seen in Freud's "Wolf Man", who showed no grief on the death of his sister, but shed bitter tears at Pushkin's grave a few months later.[10]

In a dream, the ideational material becomes transformed by dream distortion in shifting from the latent to the manifest content of the dream, while it is more common for the affects to remain relatively uninfluenced by dream distortion, so that the quality of affect of the dream may be appropriate while the content is not. The driving force of the dream remains a wish of some sort, often with some direct or indirect link to childhood experiences and wishes and early emotional experiences.

Thus, again it can be seen that "the release of affect and the ideational content do not constitute the indissoluble organic unity as which we are in the habit of treating them, but that these two separate entities may be merely *soldered* together and can be detached from each other by analysis".[11] The affect in the dream shows great mobility; it can not only be detached from the ideational material, but can also be reversed such as from hate to love, can be inhibited or eliminated, or can be diminished in intensity, all of which adds to the dream's surprising manifest content. Different affects can also come together as sources of the dream, and during dream analysis may need teasing out.

Freud also points out that a dominating element in the dreamer's mind may be constituted by a mood, or "*tendency* to some affect—and this may then have a determining influence upon his dreams. A mood of this kind may arise from his experiences or thoughts during the preceding day, or its sources may be somatic. In either case it will be accompanied by the trains of thought appropriate to it."[12] The mood may be worked over, and transformed in various ways, once more so that it can be used by the dream to express a wish, the psychical motive force of the dream.

Later, Freud in his papers on "Repression"[13] and in "The Unconscious",[14] the affect is defined as the subjective transposition of the quantity of instinctual energy, whose source is *within* the organism. Besides the term "affect", he makes use of the expression "*quota of affect*" when wanting to place emphasis on the strictly economic or quantitative aspect. The quota of affect corresponds to the instinct or drive from the interior of the body in so far as the latter has become *detached* from the idea and finds expression, proportionate to its quantity, in processes which are sensed as affects.[15]

The affects are sensed or perceived consciously in terms of various degrees or quotas of pleasure and unpleasure, whose source is *within* the human subject. The affects provide a form of monitoring of the internal state of the subject, the various shifts of psychic energy, or drives, the source of which could be chemical or neurological or both.

Affects can be discharged or inhibited in various ways, as Freud and Breuer had already begun to describe in their explanation of hysterical symptoms. There can be the "free" or automatic, or "unbound" discharge shown for example by a baby, which usually elicits an appropriate response from the caregiver, so satisfying the baby's needs. Or the discharge may be "bound" or inhibited, so that there is a delay in motor discharge—necessary for the "taming" of raw affects. In pathology, such as with psychical trauma, this inhibition may go so far as to "strangle" or distort the discharge. Ideas representing the drives are retained in the unconscious, while the affect, the *quantitative* factor, is consciously perceived, usually as anxiety. What determines the *quality* of affects or feelings is "probably the amount of excitation *in a given period of time*".[16]

In Freud's later theory of anxiety,[17] affects are seen as reproductions of early experiences which orient the ego towards adaptation. The nucleus of the affect is the repetition of some earlier particular experience. Echoing his earlier work with Breuer, he described how "Affective states have become incorporated in the mind as precipitates of primaeval traumatic experiences, and when a similar situation occurs they are revived like mnemic symbols",[18] which he likens to a hysterical attack. He further distinguishes traumatic from signal anxiety. Traumatic anxiety occurs when the psyche cannot cope with a flood of anxiety producing experience, while signal anxiety consists of a release of a small quantity of affect which signals the ego to adapt to a situation of danger. So, affects linked to traumatic experiences can be overwhelming of the ego, while small amounts of affect, if under the management of the ego, can play a useful role in warding off psychological danger so that trauma is avoided.

In his last account of affect in *An Outline of Psycho-Analysis*, Freud describes the role of the ego in managing the demands of the internal drives or of external circumstances as guided by various tensions produced by internal or external stimuli. "The raising of these tensions is in general felt as *unpleasure* and their lowering as *pleasure*. It is probable, however, that what is felt as pleasure or unpleasure is not the *absolute* height of this tension but something in the rhythm of the changes in them."[19]

Gilbert Rose in his book *Between Couch and Piano* speculates from Freud's views on affect that the interplay of tension and release either generates or reflects affect, or is at least closely related to the nature of affect; and if a *dynamic of tension and release over time* lies at the heart of affect, a similar dynamic will lie at the heart of artistic endeavour, including that of music.[20] Traditional tonal organisation is bound up with drama, with shifts between dramatic tension and stability. The dramatic pattern of the music demands a resolution of rhythmic tension, a resolution that had to be combined with the need for keeping the piece moving until the end. The core phenomenon of musical movement is the tension between tones.[21] Thus, as Rose points out, Freud's theory of affect leads to the conclusion that music and affect share the *common dynamic of tension and release*.

One could then maintain that much of the emotional satisfaction in listening to tonal music comes from experiencing tonal tensions and their resolution without necessarily tying these movements to specific affects. Music then is a mirror of emotional tensions, as Suzanne Langer basically proposes.[22] However, this does not account for a vast amount of music where there are words which particularise the emotions, such as in choral music and lieder; or in many cultures where music and emotion are intimately linked, such as with the Kaluli people I described in Chapter One; or music which adds to or heightens specific emotions as in opera; or with a number of symphonic pieces, notably with Mahler, where the dramatic or narrative element also particularises the emotions. One could then say that movements of tension and release in music are a crucial element of musical expression, but cannot account substantially for the emotions both produced and evoked by music in listeners.

Psychoanalysts after Freud have added to the theory of affect. One of the most influential thinkers with regard to impact on the wider community is John Bowlby, with his seminal work on attachment, with an emphasis on the survival value of feelings and their communication at the early mother–infant stage. For Bowlby, feelings are states of *appraisal*, a notion to be taken up by many psychologists and other thinkers. Appraisals are intuitive evaluations of the overall state of the individual, both of their internal world and of their external environment. Bowlby, like many other analysts at that time, was part of the Independent Group, where object relations theory was being developed.[23] As a result, feelings are to be seen as part of object relations, both those within the self and in relation to others.

Other Independents continued to see affects within the context of an object relationship and not confined to a single person psychology as was the tendency with Freud and Klein. This led to the realisation that the analyst's countertransference feelings were a guide to what was taking place between analyst and patient, indeed that "… the analyst's emotional response to his patient within the analytic situation represents one of his most important tools for his work."[24] The analyst's emotional responsiveness is an instrument of research into the patient's unconscious. The aim of the analyst's own analysis is not to become an emotional zombie but to be able to sustain complex emotional states without discharge of affect. Along with freely and evenly hovering attention which enables the analyst to listen simultaneously on many levels, rather one might say as if listening to a complex piece of music, the analyst "… needs a freely roused emotional sensibility so as to perceive and follow closely his patient's emotional movements and unconscious phantasies. By comparing the feelings roused in himself with the content of his patient's associations and the qualities of his mood and behaviour, the analyst has the means for checking whether he has understood or failed to understand the patient."[25] I described in the first chapter how Francis Grier used his countertransference emotional responses around issues of musicality in order to help his patient shift in her rigid way of relating. This shows not only how important it is to be working with the feelings of both analyst and patient, but also how this work can be complex and demanding.

More recent work with borderline patients has revealed even more complexity in the structure and function of affects in the life of the subject, and in the way that they may attach to, or be separate from, ideas. For example, Joyce McDougall has described what she calls a "dis-affected" patient, where affective life appears to be devoid of life.[26] Although such patients may seek help for a variety of reasons, they have one personality feature in common— they appear to be pragmatic and factual, unimaginative and unemotional in the face of important events and in their personal relationships. There is what she calls a "psychic gap" between emotions and their mental represen- tations,[27] which leads to an affectless way of experiencing events and people. Though this may reflect an ironclad defensive structure, the patients have little conscious knowledge of what they are doing. This is observable in analysis, when such patients avoid any possibility of emotional arousal. They may use various *actions* such as drinking excessive alcohol, becoming involved in frenetic sexual activity and/or drug abuse, or vomiting, as a way of dispersing their emotions. These actions all represent compulsive ways of avoiding the flooding of affects, often with the use of psychotic defences, though the patients are not psychotic. Thus, one can see in this case how action can disperse or fragment emotions rather than be a manifestation of their expression.

These patients may have had a family history where early on the mother has been out of touch with the baby's emotional needs, while at the same time controlling the baby's responses by discouraging the baby's spontaneous responses. The adults themselves have little sense of an internalised stable caregiver figure, and therefore find the psychoanalytic setting both a challenge to their habitual way of relating, and also the only safe setting where they can have a chance of connecting up with their affects in a meaningful way.

Work with such patients is very challenging and involves slow piecemeal analysis of the way that emotions are constantly avoided in any small interaction, often by the use of the analyst's countertransference, where affects can, paradoxically, be experienced, at least to some extent. This kind of work also shows how the important role that the affect of anxiety plays in how a person regulates their emotional life. The dis-affected patients may either have never had the means to regulate their anxiety, or perhaps have to live with such overwhelming anxiety whenever faced with the demands of a human relationship that they deaden their capacity to feel.

The French psychosomatic school point out that with patients suffering from severe psychosomatic conditions, there is often a difficulty in elaboration of fantasies and imagination, with instead the predominance of a particular kind of thinking—"*operational thinking*", which refers to ideation with no essential connection with fantasy and affect. As with dis-affected patients, the subject remains constantly at the level of actions, sticking closely to the materiality of facts.[28] Yet their body expresses their affects through concrete means, through the psychosomatic symptom.

In general, one can say that the management of anxiety as well as basic feeling-states play particularly significant roles in human development. As Anne-Marie and Joseph Sandler describe with regard to feeling states: "An experience only has or retains meaning for the child if it is linked with feeling. The assumption is made that ultimately all meaning is developmentally and functionally related to states of feeling, and that an experience which does not have some relation to a feeling-state has no psychological significance for the individual at all."[29]

Furthermore, in relation to the early months of life, they add that:

> We can speculate that, in the beginning, the child will have two great classes of subjective experience—those experiences which are pleasant, gratifying, comfortable and associated with safety on the one hand, and those which are unpleasant, uncomfortable and painful, on the other. The child naturally reacts to experiences which are either pleasurable or unpleasurable. If he is confronted with a situation which he perceives, in his own primitive way, as a pleasant one, associated with pleasant feelings, he will respond to it by joyful gurgling, and by other signs of happiness. If he is subjected to a situation which is painful or unpleasant, he will show a response of withdrawal, distress or both. If we speak only in terms of recognition rather than of remembering, we could postulate that the first important distinction *recognized* by the child in the world of his experience is the difference between the two basic classes of affects, of feeling-experiences.[30]

For Melanie Klein, the affect of *anxiety* remains a cardinal affect for understanding the early and later stages of human development. Thus:

> All aspects of development contribute towards the process of modifying anxiety and, therefore, the vicissitudes of anxiety can only be understood in their interaction with all developmental factors. For instance, acquiring physical skills; play activities;

the development of speech and intellectual progress in general; habits of cleanliness; the growth of sublimations; the widening of the range of object-relations; the progress in the child's libidinal organization—all these achievements are inextricably interwoven ... with the vicissitudes of anxiety and the defences evolved against it.[31]

Unconscious phantasies with Klein come to organise affects or structure primitive feelings from very early on in a baby's life, but remain active throughout life; the adult world has its roots for her in infancy. Phantasy is the mental corollary of the drives; it represents the particular content of urges and feelings, wishes, desire, and anxieties.[32]

André Green has written extensively on the psychoanalytic theory of affects, and the complex way that affects accompany, define, and qualify mental states. Being closely linked with biological drives, affects can have at times a forceful and disorganising power, when they appear suddenly or spread, accumulate, or pervade. The "*essence of affect is its dynamic attribute, its capacity to seep into other domains and inhabit them and finally to transform both itself and its products of the area of the mind which it has occupied*".[33] This transformative process is essential to the inner structure of affects.

> The core of all affective phenomena ... [is a] force capable of invading any or all parts of an individual, crossing even the borders of individuality to reach other entities, for instance, groups; bearing an impressive capacity for unpredictable change, either being influenced by the structures it inhabits or transforming them, shifting from an inside polarity to an outside one, and also consequently fixing itself either on the subject or being directed towards objects, forming the basis of the awareness of existence. This force, which is less susceptible to control than any other source of information, will have to meet two other psychical events which will shape its fate: first a basic principle of structuralization, second the effects of a counter-force as a consequence of that same basic principle.[34]

Green points to both the cooperation between affects and representations and their antagonism, as already prefigured by Freud. Affects and representations are connected with one another but relate differently to the mental organisation of things or words. While both affects and representations can be unconscious, they remain so in different ways. As I mentioned, affects tend to diffuse and have *force*, and also may involve counter-forces, that is, they may produce ambivalence, but they are not organised, or articulated like representations; they do not condense or split or displace one another as do ideas.

Affects also give tone and colour to representations. "Affects are needed to keep our minds in motion in search of new and promising experiences, but we are threatened by diffusions of affect that overwhelm us from time to time."[35] The example of the dis-affected patient as described by Joyce McDougall is an example of the potential overwhelming power of affect. It requires psychic work to manage affects, separating out the representations from the contradictory and disorganising affective infiltrations. But we also need affects, which not only keep the organism in motion through their dynamism, but also help us seek new experiences.[36]

Affects and representations can also cooperate less conflictually, and one can see this when listening to music, or participating in musical perfor-mances; indeed, I would suggest that music seems to be capable of *fusing* affects and ideas in a powerful and highly pleasurable way, intimately linking body and mind. Clues as to how this may occur at a physical level may be found in some recent neuroscientific research which emphasises the cardinal role of primordial affects in the constitution of the self, while musical understanding, with the way that music communicates states of feeling and emotion, captured by "... such kinetic terms as 'crescendo,' 'decrescendo,' 'fading,' 'exploding,' 'bursting,' elongated,' 'fleeting,' 'pulsing,' 'wavering,' 'effortful,' 'easy,' and so on",[37] can add to our understanding of how affects are shaped by experience. As I have emphasised, to fully appreciate music's impact on the emotions one needs to piece together information from a variety of disciplines.

The fact that I have just used various words to try to describe musical patterns brings up the fact that words in themselves, if only through the use of metaphor, can carry a significant emotional charge. Indeed, for Jacques Lacan, affects, being essentially intangible or phenomena which go their own way, are of little significance for his approach to psychoanalysis. This approach, much criticised by André Green, once a member of his circle, focuses on what can be spoken about, with little place for the crucial role of affects. In Lacan's view, the unconscious is organised, and is not a vague, disorganised mass of drives; it is organised in the form of a questioning, or what he called an "interrogative voice",[38] with a logic of its own. According to Lacan, psychoanalysis does not deal with feelings as such, but with a questioning of emotional states, that is, it is concerned with their meaning, in so far as they are represented in the unconscious. Hence slips of the tongue, puns, and word games are not a distraction from analytic work but are integral to it. This does leave out a vast area of early experience where words are not yet formed, and where communicative musicality comes into play. It would also seem difficult in this model to place the role of music in the subject's life. However, granted that words in themselves carry a charge both of significance and of emotion, one can understand how words with music can have an enhanced impact, as song utilises two sources of power simultaneously.

Neuroscience and musical emotion

There is a vast amount of research on different aspects of music and the brain, for example looking at musical elements such as pitch perception and rhythm and the sharing of neurological pathways between music and language, some of which I have mentioned, but my main focus is what contributions neuroscience can make to our understanding of music's emotional power, particularly around the basic role of subcortical systems in the registering and expressing of emotions and feelings.

Antonio Damasio was one of the experimental scientists instrumental in making the neuroscientific study of emotions a valid and vital form of research, counteracting the predominant tendency in previously established scientific attitudes to focus on the brain's intellectual functioning, as if the mind were a mere computer. He proposed that the mind–body split and separation of reason and emotion is untenable; instead, reason is grounded in emotion and emotion is grounded in the body. He also held that reason without emotion lacks relevance and hence is an inaccurate predictor of behaviour. His work is wide-ranging, including more recently excursions into the mind and its relation to culture. He has also, along with the neuroscientist Jaak Panksepp and the psychoanalyst and neuropsychologist Mark Solms, taken account of psychoanalytic thinking. Damasio, incidentally, as I shall cite, quite often uses musical metaphors when discussing neurological functions, and has also written on music and the brain.[39]

Damasio places feelings arising out of the brainstem at the heart of his work on the brain's functioning and the sense of self. The self is conceived as a process of states rather than a single and static structure. It arises from different elements, from the basic *"protoself"* which foreshadows the self to be, through to the core self, and then on to the autobiographical self, organised at higher cerebral levels. The higher regions of the brain are not essential for arousing or sustaining our affective experiences, though the lower levels are essential; without the latter we can have no consciousness. Evidence for this state of affairs comes from a variety of sources, including from brain damaged patients, brain stimulation, and also the existence of a rare condition, hydranencephaly, in which the cerebral cortex is destroyed in utero, usually due to a blocking of the essential cerebral circulation with accompanying death of brain tissue. Though the children with this latter condition have no functioning cerebral cortex their lower and midbrain regions are generally intact, and they show clear emotional responsiveness and learning. They are also clearly conscious beings and can express a full range of instinctual emotions. Thus, affective consciousness can be seen to be an intrinsic function of the *lower* brain regions. This really turns on its head previous notions of consciousness as essentially deriving from the higher levels of the brain, even if in this model the brain needed the lower levels to maintain basic life functions.

Conscious minds are understood as resulting from the smooth articulated operations of a number of different brain sites, including the brainstem, midbrain, and cortex, but,

> ... not in one site in particular, much as the performance of a symphonic piece does not come from the work of a single musician or even from a whole section of an orchestra. The oddest thing about the upper reaches of consciousness performance is the conspicuous absence of a conductor *before* the performance begins, although as the performance unfolds, a conductor comes into being. For all intents and purposes, a conductor is now leading the orchestra, although the performance has created the conductor—the self—not the other way around. The conductor is cobbled together by feelings and by a narrative brain device ...[40]

The protoself is attached to the body, and its first and most elementary product is what he calls *"primordial feelings"*, elementary feelings of existence, which occur spontaneously and continuously whenever one is awake, and which arise in lower parts of the brain, and are also the mental expressions of basic life support or homeostatic mechanisms within the interior of the body and which keep the body in a constant and regulated condition.

"These *primordial feelings* reflect the current state of the body along varied dimensions, for example along the scale that ranges from pleasure to pain, and they originate at the level of the brainstem rather than the cerebral cortex. All feelings of emotion are complex musical variations on primordial feelings."[41]

He also divides these feelings into spontaneous and provoked feelings.[42] Spontaneous feelings are those involved with homeostatis, and arise from the background flow of life processes within the organism.

> Spontaneous feelings signify the overall state of life regulation of an organism as good, bad, or in between. Such feelings apprise their respective minds of the ongoing state of homeostasis ... It is their business to "mind" homeostasis, literally. Feeling homeostatic feelings corresponds to listening to the never-ending background music of life, the continuous execution of life's score, complete with changes of pace and rhythm and key, not to mention volume. We are tuned to the working of the interior when we experience homeostatic feelings ...[43]

Provoked feelings are more complex and are caused by a variety of emotive responses caused by sensory stimuli and by the engagement of drives, motivations, and what he calls "emotions in the conventional sense".[44] This terminology can be confusing, but the fundamental point is that more complex emotional states ride on top of the basic spontaneous feelings which are hooked up to homeostatic processes.

Damasio does not underestimate the complexity of how the brain, starting with primordial feelings, then builds up a conscious self. "Building a mind capable of encompassing one's lived past and anticipated future, along with the lives of others, added to the fabric and a capacity for reflection to boot, resembles the execution of a symphony of Mahlerian proportions ... The grand symphonic piece that is consciousness encompasses the foundational contributions of the brainstem, forever hitched to the body, and the wider-than-the-sky imagery created in the cooperation of cerebral cortex and subcortical structures, all harmoniously stitched together ..."[45]

Images enter our mind both via the senses and from within the organism, and create various kinds of emotive responses. As well as listening to the basic "background music of life", once thoughts and words come into play one becomes "tuned" in to these thoughts rather than the world of the body, and we have a whole set of more complex emotional responses. "Emotional feelings are about hearing occasional songs and sometimes full-scale opera arias. The pieces are still executed by the same ensembles in the same hall— the body—and against the same background: life."[46] We thus become more attuned to our ongoing thoughts and feel their emotional reactions, which are a mix both of variable reactions, depending upon our own personality and life experiences, and also those arising from a standard emotional repertoire.

It is difficult to make direct connections between psychoanalytic and neuroscientific findings, though Mark Solms has endeavoured to do so.[47] As he points out, Freud thought that the excitation and inhibition of affects, in particular those of pleasure and unpleasure, regulated the mental apparatus. Affect assigns value to the state of the psyche by registering its biological consequences in consciousness, that is, affects register, or perceive, the internal state of the subject. They are seen as *perceptions* of the varying or oscillating internal states of the mental apparatus, the instinctual tensions within the person, and they are felt *directly*. Affects are the qualitative form in which these tensions become conscious. In that sense, Freud's picture of affects mirrors some of Damasio's picture of the basic role of primordial feelings in the construction of the self and in the creation of consciousness.

Affects are *discharged* either internally, altering the state of the body, or externally through action, the two forms of discharge often being indistin-guishable. Again, this matches Damasio's model of emotions involving both internal bodily processes and action on the environment.

For Freud, affects can be modified by the ego, which can delay motor discharge of affect, for example by means of thought. However, affects arising from repressed ideas, which are excluded from the orbit of the ego, cannot be modified or "tamed", and they play an important part in the development of psychopathology.[48] In addition, I described earlier how psychoanalytic experience reveals the ways that ideas and affects have different pathways, but can be soldered together, even if at times this soldering remains unstable. This has some parallels with Damasio's model of the separation of primordial feelings arising from the brainstem, and ideation from the cortex, during the construction of the self. One can only speculate about at what point, or points, in the brain such welding together and separation of affects and ideas take place.

Damasio's paper on music, feelings, and the brain, written in collaboration with Assal Habibi, is based upon his model of emotional life and its link to basic homeostatic mechanisms in the brain. They note that music of varied kinds consistently triggers a broad range of emotions and feelings, often accompanied by physiological and behavioural changes, such as changes in the state of the autonomic nervous system, with increases of pulse and respiration rates. The feelings are often pleasurable, though can also be unpleasurable with dissonant sounds, but nonetheless often contribute to overall well-being. Furthermore:

> Neuroimaging and electrophysiological studies, in normal individuals as well as in patients with focal neurological lesions, reveal that music can change the state of large-scale neural systems of the human brain. The changes are not confined to brain sectors related to auditory and motor processing; they also occur in regions related to the regulation of life processes (homeostasis), including those related to emotions and feelings, most prominently in the insula and cingulate cortices, in the ventral striatum, in the amygdala, and in certain upper brainstem nuclei. The ease with which music leads to feelings, the predictability with which it does so, the fact that human beings of many cultures actively seek and consume music, and the evidence that early humans engaged in music practices lead us to hypothesize that music has long had a consistent relation to the neural devices of human life regulation.[49]

They further hypothesise that the close relationship between music and feelings along with music's effectiveness in personal and social contexts, with its roles in homeostasis, explain, at least in part, the considerable degree of selection and replication of music-related phenomena, both biologically and culturally. They also recognise that in time the practices and uses of music became less closely aligned with its affective and homeostatic aspects and, to a certain degree, gained autonomy relative to those aspects, which may account for the varied panorama of music invention, practice, and consumption that can be found today.

Music-induced feelings thus derive from a variety of sources, both from the instinctual emotional core of the brain, and from higher brain systems which come into play when more sophisticated musical appreciation takes place. But without the fundamental role of the arousal of basic emotional states in the protoself or core self, associated with brainstem neural processing, and also probably linked simultaneously with the release of various hormones dealing with the regulation of body states as well as from the dopamine reward centres of the brain, it is unlikely that the higher cognitive processes could process music alone.[50] Music's basic emotional power, then, from the neuroscientific viewpoint, owes a considerable amount to subcortical emotional systems, which give access to primordial aspects of ourselves which otherwise would not be easily accessible.

More recent work summarised by Stefan Koelsch[51] suggests that the auditory cortex has emotion-specific functional connections with a broad range of basic emotional circuits of the brain as well as neocortical structures, and that the role of the auditory cortex in emotion is more extensive than previously believed. At the same time, music-evoked pleasure is associated with the activation of the phylogenetically old reward network that functions to ensure the survival of the individual and the species, the dopamine pathways. Yet music triggers not only basic arousal linked to these pathways, but also engagement in social functions; hence musical activity is directly related to the fulfilment of basic human needs, such as communication, cooperation, and social attachment. Supporting social functions was probably an important adaptive function of music in the evolution of humans, as I have already described in the previous chapter. Koelsch points out that the hippocampus plays a part in the generation of attachment-related emotions and can be activated by music owing to music's ability to support social attachment.[52] Of course, if this is the case, it links up with communicative musicality, the way that music is intimately related to early infant attachment to the caregiver.

There has been considerable attention paid by neuroscientists and cognitive psychologists to a particular relatively transient musical phenomenon, the experience of "chills" or goose bumps, or shivers down the spine, or "frissons", that many, though not all, people get when listening to some pieces of music, and is often felt as a moment of intense pleasure. According to John Sloboda, the frisson response is correlated with two main conditions—loud passages and passages that contain some violation of expectation, such as when there is a sudden abrupt modulation of key, or an abrupt chromatic chord or a sudden onset of a musical phrase.[53] The frisson response turns out to be a reliable one, and that even after listening to a same passage of music many times the strong emotional response recurs. According to David Huron, frisson appears to be more easily provoked when the ambient temperature is low, such as in an air-conditioned concert hall.[54]

Because the frisson response is both limited in time course and repeatable it is eminently suitable for scientific research, though one must be cautious about how relevant the response is when looking at the whole musical experience.

Huron suggests that the loudness and surprise activate the brain's fast-track fear system involving the amygdala, quickly followed by slower cortical appraisal that the situation is in fact safe, with the consequent inhibition of the amygdala. The magnitude of the contrast between the fear and the subsequent inhibition of fear produces the overall feeling of pleasure or thrill.

Neuroimaging studies by Anne Blood and Robert Zatorre have shown that several brain areas show activity that correlates with the frisson effect—those areas involved with the brain's reward system, such as the nucleus accumbens, the ventral striatum, and the ventromedial orbito-frontal cortex.[55] In addition, studies have also suggested that dopamine release is linked to the *anticipation* of pleasure rather than its fulfilment; other neurotransmitters such as opioids seem to be linked to peak pleasurable experiences.[56]

Whatever the exact nature of the neural connections, be they active or inhibited during the frisson experience, the point is that there is precise evidence that at least some aspects of powerful musical enjoyment use fairly precise and measurable brain pathways.

It is worth noting here that for Theodor Reik the experience of *surprise* is the hallmark of psychoanalysis.[57] In moments of surprise, or even shock, we come up against the eruption of the unconscious into consciousness.

> A patient is always surprised when he is told something which uncon-sciously he already knows. That follows from our designation of surprise as the reaction to the fulfilment of an unconscious expectation. He will take in what was formerly known to him and has only been alienated, as if it were something new, and will repel it. We may now understand the effect, which often does not appear till later, of such a surprising communication upon the repressed content, if we take it as a kind of psychical shock that needs time to master.[58]

For both analyst and patient, the experience of surprise represents a moment when something startlingly new from the unconscious erupts and disturbs consciousness; the content may be bizarre, unexpected, absurd, or amusing, but what becomes truly startling may well lead on to what turns out to be startlingly true.[59]

The existence of the frisson experience is a special case of the role of musical expectations in the listener's experience of music, as originally put forward by Leonard Meyer in his classic book *Emotion and Meaning in Music*.[60] He uses the psychoanalytically influenced theory of emotion advanced by John MacCurdy in the 1920s,[61] which proposed that when an instinctual reaction is inhibited or held up, then an affect is produced. This is of course what I have already described in outlining the psychoanalytic approach to affects. Meyer adds that there is also considerable pleasure or excitement in deviating from what is expected from a basic schema, norm, or mode of thought. Meyer applies his arousal hypotheses to the musical field, for example showing how moments of expectation, suspense, and eruption of something unexpected in music evoke emotional responses similar to those aroused in other life situations, so that listening is moulded by the listener's expectations.

Meyer proposed that there is a close relationship between expectations roused by music in various ways and our emotional response to music, depending upon how the composer shapes the pattern of expectations. Meyer laid particular emphasis on those expectations which are roused but are then not fulfilled, or are inhibited or blocked, which then trigger a strong emotional arousal, rather like the frisson experience but often lasting longer. Indeed, he considered the frustration of expectations was the basis for musical meaning. Thus, "... the inhibition of a tendency to respond, or on the conscious level, the frustration of an expectation ... [is] the basis of the affective and the intellectual aesthetic response to music. If this hypothesis is correct, then an analysis of the process of expectations is clearly a prerequisite for the understanding of how musical meaning, whether affective or aesthetic, arises in any particular instance".[62]

Meyer illustrates his hypothesis with musical examples, such as from the fifth movement of Beethoven's String Quartet Op. 131. He shows how Beethoven arouses the listener's expectations by a number of means, for example by introducing the theme but only incompletely melodically and harmonically, so one is left to wonder what will happen, whether or not the theme will be completed and how; one may be left at the brink of chaos, particularly when Beethoven continues with incomplete rhythms coupled with uncertain harmonies. Then just "… at the very moment when rhythm, harmony, texture, and even melody in the sense of pattern seem all but destroyed, the little figure which opens the movement and the first phrase raises our hopes and redirects our expectations of completion and return. Now we are certain as to what is coming."[63]

Furthermore, Meyer suggests, such techniques, challenging musical structure and texture, are often used whereby "… some of the greatest music is great precisely because the composer has not feared to let his music tremble on the brink of chaos, thus inspiring the listener's awe, apprehension, and anxiety and, at the same time, exciting his emotions and his intellect."[64]

One can then ask if musical meaning merely consists of these sorts of shift in the musical resources, or on the other hand if these shifts are just techniques which the skilful composer in the Western art tradition will use but are not in themselves sufficient to explain what the music means or what is being communicated, much as an actor will use various voice techniques to enhance the meaning and musicality of a play's words, but which are not in themselves responsible for the meaning of the play and the nature of what is being acted on stage. In other words, can we only understand music in its own terms, without reference to anything outside its own sphere, or is this to restrict music to a mere play of techniques and to forget how much a musical performance is embedded in human life and involves human agency and relationships? In addition, it would seem to be very restricting to consider that the main way that music affects us is through the raising and inhibition of expectancy, when much of the power and pleasure of music comes from the sheer enjoyment of melody as well as the way that music involves intense social connectedness. One can certainly understand that some basic emotional states arise when expectancy is aroused, but that more complex and subtle emotions and states of mind engage the listener in a whole musical journey.

Thus, I think one should add here in contrast that, while emotions can be aroused by expectations that are raised and then thwarted, they can also be brought out by expectations that are *fulfilled*. For example, when one hears the opening of Bach's *St John Passion*, as John Eliot Gardiner describes, there is conveyed a complex mix of light and dark emotions, the lifting up of Christ on the cross and his abasement, being brought low for the sake of humankind. In order to convey these emotions directly, there is the relentless pulsation of the bass line, the persistent sighing figures in the violas and the swirling motion in the violins suggestive of turmoil and even the surging of the crowd, and then over this ferment there is the lyrical dialogue of oboes and flutes which however produce anguished dissonances, suggested of the harrowing of Christ's flesh. Then with the entry of the chorus, there is a powerful portrayal of Christ in his majesty, looking down as it were on the "... maelstrom of distressed unregenerate humanity below".[65]

It would be difficult to explain such rich music in terms of inhibited expectations, when one is plunged straight into the musical drama.

However, there may be times when this approach can be applied, when expectations are both thwarted and then fulfilled or conflicts are resolved. Thus, Meyer describes how at the beginning of Beethoven's *Ninth Symphony*, a "progressive weakening of texture, together with both harmonic and motivic incompleteness and ambiguity, create powerful expectations whose inhibition and ultimate resolution into a clearly defined theme produce a powerful affective experience ... The sense of the relentless power of inexorable fate which characterises the main theme of the first movement is a result not only of the elemental force of the theme itself and of the ambiguities and expectations excited by the introduction but also of the particular manner in which the introduction leads into the theme."[66]

One could add here that, as important as arousal mechanisms may be in sparking off emotion, one can see how interest is maintained by the ability of the music to *sustain* emotion. For example, listening to that opening of Bach's *St John Passion* or the Kyrie of his B Minor Mass, there is an *ongoing* sense of awe and pleasure, contemplation, and sheer excitement, lasting minutes. One may speculate once more that music's capacity to sustain ongoing communal excitement had a strong evolutionary power to sustain communities.

Music of course does not need the human voice to create this effect; it only needs the dynamic play of the various instrumental voices working together at multiple levels to create ongoing musical interest, from a string quartet to a symphony orchestra. There is a kind of parallel between language and music here. As I have suggested,[67] the conscious thread of thought has many streams or voices or drafts. Through dialogue the themes of these voices may become more or less coherent, and in an analytic session, one may be able to tease out the different voices and their origins. Music by its very nature of being able to manage the expression of multiple voices simultaneously has even more capacity than language to convey complexity.

This may also mean that music can express not only various forms of sonic dissonance through musical means, but may be able to help us tolerate dissonant streams of experience. Part of the power and the pleasure of listening to music, and which may help to account for why music can relax us at tense moments, is not only the sheer delight of melody but the way that conflicting streams of music can cohere and produce resolution. This may be incredibly pleasurable but also relaxing and stress reducing at some deep level.

Another take on the way that music functions at a neurological level comes from looking at *systems* rather than individual neural pathways or brain regions. The cognitive psychologist William Benzon, in his somewhat speculative but fascinating book *Beethoven's Anvil*,[68] suggests that music is a powerful means of coupling nervous systems in social interaction; that music connects us to the social world and allows us for a while to radically reconceive and reconstruct our relationship with the world. He sees people's brains as intimately involved with the social world rather than as isolated from others.

He begins by challenging Cartesian thinking, with its split between mind and body, reason and emotion, and its emphasis on the lone individual isolated from other individuals. To make his case, he starts from Damasio's evidence from neuroscience that the mind–body split and separation of reason and emotion is untenable; instead reason is grounded in emotion and emotion is grounded in the body, from which he moves to how people acting together to express emotion sustains human culture, and is the basis for the musical experience.

Benzon argues that coupled nervous systems in some sense function as a single system, and that music is a medium through which individual brains are coupled together in shared activity.[69] Humans create social spaces where their nervous systems are coupled through interactional synchrony, what I had earlier described as entrainment. Musicking is adept at this coupling as it uses neural circuits at all levels of the brain. Music is very much an exercise in timing, and Benzon hypothesises that it may serve as means of coordinating the temporal structures of widely distributed brain regions.[70] If, Benzon, argues, Damasio is correct to think of our emotional life as being grounded in the neural regulation of the state of the body, then it

> … follows that spontaneous eruptions of feeling reflect the activity of
> those parts of the brain concerned with monitoring our bodily state

and with communicating that state to others. The latter is crucial. We are social creatures ... When we express emotion we are signalling something about our internal milieu. We assume others will pick up the signal and respond accordingly.[71]

One can see how this happens in for example a group of workers singing together, where in the act of singing the workers link "... their minds and bodies into a single dynamic system, a community of sympathy. By bringing their work into that same dynamic field, they incorporate it into that form of *society* created through synchronization of interacting brains."[72]

He further argues,[73] that the neural mechanisms for music include a repetitive, or "groove" stream, which carries the pulse of musical performance and is mediated by the locomotion systems, such as the basal ganglia and cerebellum as well as at the cortical level, and is extremely precise. The other stream is the "gesture" stream which creates the meaningful "story" of the music. The latter is organised by limbic structures in the hippocampus and the adjoining entorhinal cortex responsible for the brain's navigation system. "When used for music, the navigation system is linked to the various cortical regions supporting the recognition and manipulation of musical sound—regions for recognizing intervals, melodic contours, harmonic relations, and tone quality. These regions implement the musical 'space' through which the [gesture] stream navigates."[74]

One could speculate that music has the unique power of uniting the groove and gesture streams when people are interacting socially.

There are some similarities between these ideas and Trevarthen's notion, which I mentioned in Chapter Two, that humans move under the coordinated and integrated control of a time keeping energy-regulating "intrinsic motive pulse" (IMP), originating in the brainstem. The IMP is conceived as a series of generators of neural and body-moving time, which forms part of a larger *system* of generators regulating our emotions, movements, and thoughts.

Patrik Juslin and colleagues have provided a comprehensive theory of musical emotions, putting together findings from neuroscience and cognitive psychology.[75] This theory proposes eight distinct ways by which music can arouse emotion in listeners, the so-called BRECVEMA framework. Some of these mechanisms are instinctive emotional responses, hard-wired into the brain though modified by experience, and others depend more upon experience and learning. The assumption behind the theory is that emotions are adaptive because of their ability to evaluate what is happening in the environment, and that mechanisms of emotion induction are, "*... infor-mation-processing devices at different levels of the brain, which utilize distinct types of information to guide future behaviour.*"[76]

These mechanisms in summary are the following:

1. *Brainstem reflexes.* These are quick, automatic, and unlearned responses to a potentially important or urgent sound needing attention. A sudden loud sound or change in pitch in music may increase arousal and evoke feelings of surprise in the audience, such as the kettle drum stroke in Haydn's *Symphony No. 94*, the so-called "*Surprise Symphony*".

2. *Rhythmic entrainment.* This is when an emotion is evoked by a piece of music in which a powerful external rhythm influences an internal bodily rhythm, such as when we tap our foot or dance to the beat of the music. Entrainment both increases arousal and the sense of communion and bonding between people.

3. *Evaluative conditioning.* This is when a piece of music can induce emotions because of its being paired, as a positive or negative experience, with the previous context in which it had been heard. The use of melodic elements to evoke particular emotions associated with a character or a theme can be seen in Wagner's use of the so-called leitmotifs in *The Ring*.

4. *Emotional contagion* occurs when we mimic a piece of music internally and start to feel the same emotion by some mirroring process, probably involving the mirror-neuron systems in the brain.

5. *Visual imagery.* This is when an emotion is evoked in the listeners when they bring to mind an inner visual image while listening to the music.

6. *Episodic memory.* This is the memory of specific events in one's life that become associated with a piece of music—the "They are playing our tune" phenomenon, as seen in the film *Casablanca*.

7. *Musical expectancy.* This is when, as described above, an emotion such as frisson, surprise, or anxiety is induced when a specific piece of music violates, delays, or confirms the listener's expectations about what is going to happen next.

8. Finally, *Aesthetic response* to music is when the listener recognises and appreciates the sheer skill that has gone into a performance. This is of course very much influenced by culture and personal taste.

The notion of BRECVEMA is that all these mechanisms may well be directed at a particular musical event to contribute to the listener's overall emotional experience. The theory integrates a great deal of research information into one unified view, and in that sense is both impressive and useful as pointing towards further areas of research. Even if one may take issue with particular elements of the theory, the point is that it shows how one needs a multi-level approach to understanding musical experience. Juslin's recent book *Musical Emotions Explained*[77] provides extensive empirical evidence for his take on musical emotions. However, the approach is very much from a behavioural perspective, which still leaves open for consideration many aspects of human experience and musical life that cannot be tied down to such a tight scientific framework and for which we need other approaches to comprehend them.

Social theories of musical emotion

Another theory which emphasises the interaction between the subject and the environment in how we perceive music derives from that of the ecological approach to perception first devised by James Gibson.[78] This is an approach which emphasises various invariant elements in the environment as providing opportunities or "*affordances*" for the subject to respond to or pick up on. It implies the complementarity of the subject and the environment. A chair affords the opportunity for an adult to sit down, or a child to crawl under. Affordance was coined by Gibson "to stand for the opportunities, functions, and values that a perceiver detects in the environment, arising from the mutual relationship between the needs and capacities of the organism, and the properties of objects and events".[79] Music affords the opportunity for dancing, contemplating, entraining groups together, and so on, as well as affording certain kinds of interpretation and not others. As Eric Clarke points out, the recapitulation in the first movement of Beethoven's *Ninth Symphony* has been described by Susan McClary as showing murderous sexual rage, or by Donald Tovey as the heavens on fire, but no one could think it revealed world-weary indifference; the last interpretation does not match with the music's material properties or semantic requirements. The musical material, then, shapes the character of a listener's response or engagement, or it influences their "subject position", or attitude to what they are listening to.[80] One could then see pieces of music as objects available for transforming the subject's emotions; music is one of the means by which the subject can regulate emotions.

Tia DeNora is less concerned with the mechanisms by which music affects us and more interested in how music's power is linked to the way that it is used in social situations, by using and interpreting data from a series of ethnographic investigations of music in daily life, including interviews with women in the UK and US in a variety of settings, such as in exercise classes, karaoke evenings, music therapy sessions, and in the retail sector where music is used to influence customer choice. The aim of this research was to see "how music articulates social life and social life articulates music."[81]

She cites as an exemplar of this form of study the classical work of young people and their intimate involvement with music in Paul Willis' book *Profane Culture*.[82] In this work, the culture of the "bikeboys" is examined, with their preference for music which has a strong beat and a pulsating rhythm. Their preference was for music which did not allow for passivity, but incited movement; and if they couldn't dance anymore they'd then be driven to go back on their bikes. The music and their lifestyle were intimately connected. As DeNora describes, this setting shows "music as active and dynamic, as constitutive not merely of values but of trajectories and styles of conduct in real time. It reminded us of how we do things to music—dance and ride in the case of the bikeboys, but beyond this, work, eat, fall asleep, dance, romance, daydream, exercise, celebrate, protest, worship, mediate, and procreate with music playing."[83] Music as a medium of social relations acts directly through body and mind and has the power to shape human agency.

Through the evidence of the extensive interviews, DeNora charts how music helps to regulate emotions at a bodily level, can shift people's moods, enhancing togetherness through musical entrainment, and in that sense can become a container for feelings.[84] Music can become a device for regulating and structuring social encounters, both in a positive and negative sense. While music can enhance individual and group well-being, it can also have a less benign effect, for example by influencing consumer choice in supermarkets, or by facilitating aggression through martial music. Overall, however, she sees music as having the profound power to make available ways of feeling, being, moving, and thinking, that it animates us, that it keeps us "awake".[85]

In the next chapter I will move onto other theories of music, which, like DeNora's, focus on understanding the human aspects of the musical experience, though with more emphasis on the humanities and less emphasis on empirical research.

CHAPTER FIVE

Music and emotion, second movement

I have been covering various theories of musical emotion throughout the book; they can be summarised under three main areas, depending upon music's power or influence over the human subject and also on the model of emotions being used to understand music.

1. Where music imitates or represents emotions.
2. Where music arouses emotions.
3. Where music expresses emotions.

However, each of these areas overlaps considerably when considering a particular theory or approach. For example, as Juslin emphasises, perception and arousal of emotion may appear to be different processes, and indeed may have different mechanisms, yet they often occur together, and may influence each other in various ways.[1]

There has also been a tendency in some theoretical writings on music to deny or minimise the intimate connection between music and emotion, resisting the notion that music is about or represents human emotions and emphasising music's formal or structural properties as sufficient to explain music's effect on the listener, for example relying on the evident emotive powers inherent in various tonal tensions.

Others would see music's power to convey a complex range of feelings as evident, but these are aesthetic feelings, that is, feelings distilled by the composer into artistic form; like every art, a musical piece, if it is to be interesting and have an impact, is a complex work put together in a way *never* experienced before. The composer chooses the route and the pathways of the musical journey for us to enjoy or engage with. One may add that even music of other cultures has an element of this aesthetic function, whether or not it is produced at weddings, funerals, as part of ritual or trance, and so on. There is also the question of how much of a narrative element, with or without words, is required to channel the complexity of the emotional message, charting the unfolding of an interior drama, as one can see with Mahler's music.[2]

Music as an art form can be lost in the various elaborate discussions on whether or not music conveys specific emotions and the origin of the emotions. Music is perhaps more like poetry in that it carries information about emotions in a concentrated form. Thus, a listener can recognise sadness or joy in some music but is not necessarily made sad or joyful, though may be so briefly. If one thinks of a piece of theatre such as Shakespeare's *Othello*, or Verdi's opera *Otello*, the lead character's suicide and the events leading up to it move us, but we are not usually made to feel suicidal. Instead we have an *aesthetic* experience, a special form of emotional experience. For example, what moves us intensely is the way in which the artist makes sense of the *Othello* tragedy, welding together complex situations, unconscious or obscure or fragmentary feelings and inarticulate emotions into an artistic whole, the artistic form, in a way never expressed before. In a strange way we enjoy our experience of entering into a tragic world, perhaps because tragic feelings and experiences are made intelligible in a coherent structure in a special setting, the "other scene" of the stage, opera house, or concert hall. Of course, we must have the general capacity to have an emotional experience in order to respond emotionally to what is happening on stage or in the concert hall.

The Russian psychologist Lev Vygotsky considered the only difference between aesthetic and ordinary feeling was that aesthetic feeling is released by extremely intensified activity of the imagination.[3] This is similar to what I shall describe below as the intensity of artistic vision.

One may ask what makes the aesthetic experiences of music different from those of the other arts. According to Jerrold Levinson,[4] the principal elements of the aesthetic appreciation of music consist of, first, the appreciation of *form*; on the small scale, this refers to appreciation of melody, or movement on pitch space, and awareness of motifs and phrases, rhythm, harmony, timbre, or tone colour. On the large scale, there is the awareness of melodic repetitions, the harmonic scheme, and various balances and symmetries. The next aspect is that of the appreciation of *motion* or *movement*, which we hear in an imaginary or metaphorical space, when, for example, we hear music rising, falling, rushing, lingering, and so on. "Heard musical movement is the scaffolding for almost all that engages us in music beyond the level of the basic parameters".[5] Then there is the impression of *agency* that music evokes by virtue of the movement we hear, which can be brought under the heading of *gesture*, by analogy with the role of physical gesture in the behavioural expression of emotion. Music appears to be pervaded by various gestures, such as sighing, caressing, threatening, despairing, or rejoicing, all of which then reveal musical *expressiveness*.

Music can also reveal *non-expressive aesthetic qualities* such as higher-order perceptual qualities including grace, delicacy, charm, humour, menace, mystery; it can have *narrative-dramatic content*, and finally it can have in some cases, such as programme music, clear *representational content*.

The aesthetic appreciation of music, as Levinson emphasises,[6] involves not only the perception and recognition of all these types of musical expressiveness, but also responding emotionally to that expressiveness; being moved in some way by emotionally expressive music, at least on some occasions, is the *sine qua non* of appreciating it, he suggests.

From the performers' position, they have to find a way of conveying the meaning and emotions of the text or song without themselves necessarily experiencing the particular emotion being conveyed. As the tenor Thomas Moser put it, "You have to express emotion but you can't sing with emotion."[7] Too much emotion in the voice just clogs up the performance. That is different from the necessary state of being in touch with one's own emotional state and understanding the emotional journey that needs to be conveyed in the performance. There are also times, as the soprano Gillian Keith has pointed out, when feeling genuine emotion will be beneficial to the performance.

> If the performer is able to relate to the emotion in a very real way, the performance will surely be enhanced. If I am singing in a religious or spiritual work in a particular setting, for example a beautiful cathedral, I might feel a certain sense of awe and admiration coming into my voice as I am being moved by the music in these particular surroundings. I believe that one would be able to hear these things in my voice as my singing was lifted by the other-worldliness and the spirituality of the occasion.[8]

Similarly, with a composer, the act of composition does not require the composer to be in a particular state of mind, other than that of absorption in the act of composing. The composer does not have to feel a particular emotion or dramatic situation in order to convey this in the music. But they do need to have the technique to be able to know how to compose music that will convey the dramatic expression for which they are searching. It is highly unlikely that a composer could do this if they were unable to feel emotion, or know what kind of dramatic tensions they are seeking without having experienced real-life dramas. A composer, as with any artist, "has to weave the emotions he is expressing into an intellectually and emotionally coherent statement; and emotions woven together in this artistically formal way do not cease to be emotions because they do not float about vaguely as in everyday life; in fact, they become even more 'real' by their isolation and sensitive combination in a great work of art. The great artist makes a supremely 'right' statement of the emotions one feels oneself but cannot organize into a satisfying experience".[9]

There are some pieces of music, notably so-called "programme music" in which the emotional information conveyed is more obviously representational or quasi-representational. For example, Tchaikovsky's fantasy overture, *The Tempest* (Opus 18), reflects Shakespeare's play. There is the opening calm and then the raging storm, the love theme, and final calm. There is a complex mix of narrative, stormy emotion, and subtle musical form, which conveys something essential and *new* about Shakespeare's story, without representing exactly the plot and poetry of the play. Yet as with any artistic medium, we experience a concentrated vision of life.

Lawrence Kramer[10] also links the musical experience to feelings of strangeness and the uncanny, and hence close to unconscious experience. Uncanny experiences include those that are frightening and arouse a sense of horror and dread. Freud in "The Uncanny" traces such experiences back to what is previously known and familiar, and yet which erupt in unexpected ways. The title *Das Unheimliche* in German can be traced back to what is homelike, what belongs to the house, but also something that becomes concealed.[11]

Typical uncanny experiences include inanimate objects apparently coming to life, a sudden appearance of a double, the appearance of ghosts and spirits and other hauntings. Something becomes uncanny when the distinction between imagination and reality is effaced. Ultimately, the uncanny is something which is secretly familiar and has undergone repression and then returned from it—hence the double feeling of the strange and the familiar that is indicative of an uncanny experience and which links such experiences to irruptions from the unconscious.

Kramer describes how music reaches directly and deeply into the unconscious, hence its elusiveness to being tied down by language. In order to make sense of music, we have to rely on metaphor, allusion, indirect descriptions, or else a too hermetic musical analysis in the hope that this will tie down its meaning. But that meaning cannot often be tied down, by its very nature. However, one may add that once there are words linked to music, the meaning of the music can become easier to follow, even if it is open-ended. Thus, to give one example—Schubert's song *Erlkönig*, "The Erlking". A boy is carried through a forest with his father on horseback, terrified and entranced by the words and song of the demonic Erlking, who tried to lure the boy to his dark world with apparently sweet tones. The song is based on a Goethe poem, and has a dream-like quality. The musical figure, the hammering repeated octaves, represent metaphorically the horse's hooves, the heartbeat of the father, the fear of the son, and the knocking on the door of fate or death. The music conveys all these meanings, consciously and unconsciously, which gives the song its depth and its emotional trajectory. There is also the fact of the words which add some precision through linguistic meaning. However, the substantial and unconscious level is represented by the music, which conveys the multiple meanings of the boy's journey, ultimately to his death at the end of the song.

For those who perform choral music the words are more intimately integrated with the music. Thus, Harry Christophers, founder of The Sixteen, looks at the text of songs in order to get as close to it as possible; then he formulates his approach to the music based on the text.[12]

For poets, the music in the words is crucial to its impact and meaning. As John Burnside puts it, "… what matters most in a poem is its music and how it refreshes the language, strengthening it against the abuses of the unscrupulous and the careless, and allowing it to retain its ability to enchant, to invoke and to particularize."[13]

Kramer focuses on the specifics of the musical experience when tackling the nature of the emotional power of music.

> When we say that music "expresses" emotion we don't mean either
> that it signifies emotion (it's more immediate than that) or that it
> arouses emotion (it does, but it is scarcely unique in doing so). Music

has two specific emotional powers at which it is, if not unrivalled, unsurpassed. First, it renders emotions tangible, giving a sensuous, reproducible form to something otherwise transient and interior. And it does so without sacrificing the force and plasticity of feeling; it does not objectify, but extends subjectivity beyond the boundaries of the nominal self. Second, music detaches emotions from specific motives and circumstances, giving it an independence that is also a form of pleasure, even when the emotions are dark or disturbing. And it does so without giving an effect of abstraction; the feelings involved always seem specific, not generic.[14]

Cognitive and neuroscientific explanations of the power of music that fail to take account of unconscious experience, the subjective experience of music, and the "musical space" where we hear the music, can give us only limited understanding of the total musical experience, though findings from these disciplines are still important in providing insights into the complex processes of musical perception. And these disciplines also contribute to understanding the nature of *early* musical experience, its relation with time and the shaping of early affects, reflecting the flow of inner life, which provides the basis for understanding the fully formed musical experience. Thus, one can say that music allows us to get in touch with an early form of connection with the mother–infant relationship, however sophisticated its later use in adult life. With the latter, music can paint specific emotions and situations as with Schubert's *Erlking*, or just allow us to enjoy a mood state, or a flow of emotion without any particular object.

Rather than go systematically through theories based upon a specific model of emotion, given the often overlapping nature of many of these theories, I will select approaches covering a number of key areas, starting with what I suggest remains still a foundation for understanding the total musical experience, the Ancient Greeks' musical experience.

Music was as central to Ancient Greek life as it is to ours. Greeks believed that music had the power to captivate and enchant. In the case of the Sirens' song, it could beguile listeners to the point of death. The Ancient Greeks had a model of music as imitating emotion. But this model of "mimesis" goes beyond what we would usually understand by the term music and imitation. As Armand D'Angour points out,[15] the Muses, from which the term *mousiké* derives, include a variety of disciplines and genres, such as dance, tragic song, poetry of various kinds, and historical narrative, as well as the visual response to the performance of the songs and recitations, all embedded in the overall Greek way of life. We have evidence of a notated music from various sources—a piece of papyrus from the Orestes of Euripides from 408 BC, and stone tablets from Delphic hymns as well as some songs on stone columns. Ancient Greek music was mainly sung music; its rhythms were determined by the signs above the words, which also indicated the pattern of dance steps, and the pitch intervals of the melodies were also indicated by the signs of the alphabet—the more distant from the letter alpha, the higher the pitch. Thus, as D'Angour points out, the Ancient Greeks did have a notion of music as we understand it, embedded in the overall notion of *mousiké*, and indeed he has himself revived what it may well have sounded like, using reconstructions of both texts and instruments.

D'Angour points out that one must also add an ethical dimension when considering *mousiké*, embracing the broad realm of what the Muses represent.[16] *Mousiké* in various forms sounds beautiful, pleasing, and sweet, and conveys delight, pleasure, and enchantment, while a sung performance has the power, along with the poetry, to move the listener to joyful laughter or tears; poetic meaning and musical sound were intricately connected. But Ancient Greek thinkers who speak and write about *mousiké*, "... are less inclined to express aesthetic appreciation than to consider its moral, intellectual, or religious significance ... The extended discussions of Plato and Aristotle revolve predominantly around the different kinds of ethical character (*ethos*) associated with and imparted by different rhythms and *harmonias* (musical modes). Plato's main concern was that these have the power—through some mimetic process that he assumes but never fully explains—to affect the condition of the *psuché* [soul]."[17]

For Plato, the pleasure of music should be guided to good purpose through the use of reason, while Aristotle recognised that music could be beneficial as a means of relaxation and a source of social enjoyment; but both considered that ethical criteria should be used to guide the listener's choice of music, such as how music may improve a person's moral character. Overall, the Greek doctrine of ethos or character was linked to the conviction that different kinds of music affect character, for example by producing an uplifting and calming effect, or producing excitement and enthusiasm.

The judge of *character* was integral to the plots of Greek tragedy, where individuals' weaknesses or strengths of character determine, along with the influence of the gods, their destiny. That which makes for right action was also integral to Greek ethical thought; the understanding of music's power was embedded in this particular way of life. Thus, Aristotle discusses the role of *mousiké* in his *Politics*, and whether or not it produces education, amusement, or intellectual enjoyment. He recognises that *mousiké* is pursued not only as a natural alleviation of past toil and as providing recreation, but it has a significant influence on people's character and the soul:

> And that they are so affected is proved in many ways, and not least by the power which the songs of Olympus exercise; for beyond question they inspire enthusiasm, and enthusiasm is an emotion of the ethical part of the soul. Besides, where men hear imitations [*mimeseown*], even apart from the rhythms and tunes themselves, their feelings move in sympathy. Since then music [*mousiké*] is a pleasure, and virtue [*areté*] consists in rejecting and loving and hating aright, there is clearly nothing which we are so much concerned to acquire and to cultivate as the power of forming right judgements, and of taking delight in good dispositions [or in good characters, *ethoi*] and noble actions. Rhythm and melody supply imitations of anger and gentleness, and also of courage and temperance, and of all the qualities contrary to these, and of the other qualities of character, which hardly fall short of the actual affections, as we know from our own experience, for in listening to such strains our souls undergo a change.[18]

Such a wide and encompassing view of the effect of music on our body and soul is matched by what Richard Wagner took from Ancient Greek thought in his wish for a total work of art that would express human beings in their totality. As Bryan Magee points out,[19] Wagner saw opera as parallel with, or as the modern equivalent of, Greek drama. The latter was a composite art-form, using instrumental music, verse, song, dance, mime, and narration, all coming together on stage. The Greeks also used subject matter from mythology which encapsulated fundamental truths about human character and individual destiny, and human participation was maximised in that the dramas involved the whole of the community engaged in and witnessing the performance. This provided a model and inspiration for Wagner's own new conception of opera, making it for him equivalent for his society what Ancient Greek drama had been for its times.[20]

As one can see most vividly in Wagner's cycle of operas, *The Ring*, the orchestra would come to function for Wagner like a Greek chorus, setting the scene and commenting on the action:

> ... heightening significant moments, encouraging and rejoicing, remembering what the characters had forgotten or did not know, foretelling a future unknown to them, breaking out into lamentations or warnings and drawing it all together at the end—all functions that could now be performed better by the symphony orchestra. If a musical motive were introduced in connection with a particular character, emotion, object or situation, its subsequent use would recall that original association in the listener's mind, and thus enable the orchestra to reminisce—and equally to look forward, and to combine disparate associations, thus putting at its disposal all the resources of dramatic irony ... The potential for musical metamorphosis was protean, and also endlessly subtle.[21]

With the new music drama, thought can be conveyed through words, with all the subtleties that speech can convey, with all the clarity and potential absence of ambiguity, denied to music. While music is the natural mode of utterance of *feeling*,

> Music cannot specify in the way language can, but it can express huge emotions with overwhelming eloquence ... [I]f words and drama carry out the tasks of specifying characters and setting up situations, music will be able to speak out their emotions as nothing else can. And thus, the requirements are fulfilled for a total work of art that will express the human being in his totality of body, intellect and heart.[22]

Some of these ideas about the potential for opera to reflect the fullness of human experience had already been put forward by Arthur Schopenhauer, though Wagner was yet to discover his writings in detail when conceiving of the total art work. I have already mentioned in Chapter One that music is the voice for Schopenhauer of the metaphysical will, the inner being of the world, which is why it appears to speak to us from the most ultimate depths, deeper by far than those accessible to other arts, while remaining unamenable to language or to intellectual understanding. Music acts directly on the will itself, that is, "... feelings, passions, and emotions of the hearer, so that it quickly raises these or even alters them".[23]

Because music has this power, then in conjunction with words in opera, potentially, "... it gives the most profound, ultimate and secret information on the feeling expressed in words, or the action presented in the opera. It expresses their real and true nature, and makes us acquainted with the innermost soul of the events and occurrences, the mere cloak and body of which are presented on stage."[24]

It is worth noting that Schopenhauer also considers purely instrumental music, which can also reveal human passions and emotions in innumerable shades, but only for him in the abstract and without any particularisation, even if we clothe the work with specific emotions through our imagination.

In addition, Schopenhauer considers how music acts on the mind to produce its effects, suggesting a sort of mimetic model. Music consists, "... generally in a constant succession of chords more or less disquieting, i.e. of chords exciting desire, with chords more or less quieting and satisfying; just as the life of the heart (and the will) is a constant succession of greater or lesser disquietude, through desire or fear with composure in degrees just as varied".[25] Thus, music reflects the movements of our inner world, in particular when we have desires fulfilled or thwarted.

Going further, Schopenhauer points out that the phenomenon of musical *suspension* has particular relevance in matching human desire, and it was this text that later greatly influenced Wagner's own music. Suspension occurs where, for example, just before a final resolving chord or cadence there is a moment of dissonance which leads us to desire resolution. "It is a dissonance delaying the final consonance that is with certainty awaited; in this way the longing for it is strengthened, and its appearance affords the greater satisfaction. This is clearly an analogue of the satisfaction of the will which is enhanced through delay."[26] Then if we move from discord to another discord, instead of rest we have *surprise*—an anticipation perhaps of Meyer's theory of musical expectation.

For Wagner this description of the powerful role of suspension had a profound effect, and led him to conceive of composing a whole piece of music and indeed a whole opera in which that suspension operates. "The music would move all the way through from discord to discord in such a way that the ear was on tenterhooks throughout for a resolution that did not come. As Schopenhauer spelt out, this would be a purely musical equivalent of the unassuaged longing, craving, yearning, that is our life, that indeed is us."[27] One can of course see this process in action in *Tristan and*

Isolde, where the opening chord with its endless suspensions sets the tone for the desires and ultimately unrequited love of the protagonists.

Like Schopenhauer, the philosopher Kathleen Higgins sees music as intimately involved in our way of life and, like the Ancient Greeks, that it has an essentially ethical dimension. The theme of her book *The Music of Our Lives*[28] is that music has much to contribute to good living. Citing Plato's notion in the *Republic* that education in music is sovereign because, with the right training, more than anything else rhythm and harmony find their way to the innermost soul, imparting grace, drawing also on the Confucian notion that benevolence is akin to music, she sees music as a central tool for the promotion of harmonious living, both for the individual and their society. The current tendency to focus on issues of musical form or on one narrow aspect of musical experience has lost touch with music's wider value. "Music's capacity to engage our intellectual, emotional, and physical natures simultaneously; its suitability for promoting social cohesion; its reflection of practical and ideal modes of social interaction; its ability to stimulate reflections regarding our basic values—consider the role of music in churches—all these are basic features of musical experience. Yet these are lost to philosophical attention when 'music' is defined as 'a musical score' and aesthetics becomes a technical enterprise."[29]

She sees music as a profoundly emotional phenomenon, and that the felt sense of engagement basic to musical experience is related to our ethical capacities and orientations. "Music's affective character, which involves intersubjective empathy and often shared delight, makes listeners socially aware of their intimate connection with others—and does so in a context in which social and individual existence are not at odds. Engaging in satisfying shared experience heightens our receptivity and emotional sensitivity."[30]

As she describes, engaged and empathic listening is crucial to ethical living. Music not only facilitates attitudes towards oneself that are valuable to living well in a social world "… it also actively exercises and strengthens certain abilities that are involved in ethical action and thought. The typical stance in musical listening involves a sense of sharing a life with others. While listening, therefore, we perceive others as compatriots whose immediate interests coincide with our own."[31] Both listening to music and taking part in performing music can bring us together in a non-defensive and relatively non-competitive way; it creates intimacy and cultivates appreciation, thus facilitating mutuality and emotional entrainment. This very much matches how, as I mentioned in Chapter Two, Daniel Barenboim[32] describes that music's capacity for engaged conversation, its dialogic quality, can help in mutual understanding between people who might otherwise be deaf to what they have in common. His West–Eastern Divan orchestra, formed from Israelis, Palestinians, and Arabs, is a concrete manifestation of this hopeful principle. One can see here the power of music's ability to bring people together in a mutually satisfying endeavour, breaking down barriers to understanding and facilitating and heightening mutual communication, and hence in this context as an essentially ethical activity. Indeed, for Barenboim, the power of music lies in its ability to speak to all aspects of the human being—the animal, the emotional, the intellectual, and the spiritual.[33]

The philosopher Aaron Ridley also takes a wide view of the role of the value and place of music; for him music is a central "part of life", not to be pigeonholed into some theoretical or scientific backwater.[34] He disputes the notion that music can only or predominantly be seen and appreciated for its formal properties. He disputes the views of the philosopher of music Peter Kivy, for whom pure music is "… of the mind and of the world. It is not about the world, or about anything else, except, perhaps itself."[35] For Kivy, music, in particular pure instrumental music, is a "quasi-syntactical structure of sound understandable solely in musical terms and having no semantic or representational content, no meaning, making reference to nothing beyond itself".[36] This notion of "music alone" goes back to the nineteenth-century music critic Eduard Hanslick[37] for whom music was capable of producing the fluctuations of our inner activity, which can be similar with different feelings; but music for him could not express particular feelings. Music's beauty consists of a specifically musical kind of form—that of the combination of tones and their artistic pattern; its beauty depends not on the emotions evoked by the music in the listener, but on the objective properties of the music's composition to which one needs to listen. As I mentioned in Chapter One, the notion of "absolute" instrumental music, purified of its link to text or extra musical references, is a recent historical construct, linked to German Romanticism and the search for the "essence" of things. It also leaves out the vast realm of music from cultures across the world where music is linked to the arousal and experience of emotion.

Interestingly, despite his formalist views about musical emotion, Hanslick does appreciate that feeling is an integral part of musical performance. "To the performer it is granted to release directly the feeling which possesses him, through his instrument, and breathe into his performance the wild storms, the passionate fervour, the serene power and joy of his inwardness."[38]

Like Barenboim, Ridley sees music as connected to the way we see ourselves in the world; profound music, like profound art or thought of any kind, has the capacity to affect us at a fundamental level and perhaps to transform the ways in which we think about ourselves, the world, and our place in the world.[39] A piece of music is also profound when it enhances significantly the system in which it is embedded; it adds to the musical tradition of which it is a part, and here formal properties have their place. But a work of music, like any art, is profound when it increases our insight into our understanding and appreciation of the world, it adds perspective, and shows emotional depth; it feels truthful in some difficult to define way, and opens up horizons rather than closes them down.

Ridley illustrates the capacity of music to offer answers to deep questions about the world with a detailed account of Sibelius' late tone poem *Tapiola*, his last major work and written in 1926. *Tapiola* portrays Tapio, the animating forest spirit mentioned throughout the Finnish saga Kalevala, which Sibelius had already used as inspiration for a number of his works. Sibelius, when asked by his publishers for some illustrative text wrote:

> Wide-spread they stand, the Northland's dusky forests,
> Ancient, mysterious, brooding savage dreams;
> Within them dwells the Forest's mighty God,
> And wood-sprites in the gloom weave magic secrets.

On the one hand, it is clear that *Tapiola* works as "pure" music, in the sense of its formal properties and means of expression. It thereby creates a new world of sound and displays original handling of the orchestra, but Ridley sees its power as going well beyond these formal elements, due to its extraordinary mixture of elements and in particular in the range of its expressiveness, or what I would call its *emotional landscape*. "One hears, once its rather icy and faceless thematic material has first been announced, much that is certainly merely chilly. Yet one might also feel tempted to say that one can hear moments of relative levity, and perhaps also of awe, ardour, and, arguably, sheer terror."[40]

Ridley suggests indeed that chilliness structures the music; its title and Sibelius' own notes take us to the forests and icy wastes of northern Finland,

and in that landscape, there are passages of music which indicate not just awe and terror but sheer loneliness. Thus, "*Tapiola* is to be heard as representational, not merely of the Nordic landscape, but of that landscape as it is embodied in Tapio himself, its god ... It tells us that if we read ourselves back into nature, and back into *that* landscape, specifically, and are as unflinching as possible in our estimation of it, we conjure up a god whose heart must break; whose frozen, desolate solitude is, in the end, unbearable."[41] Of course it is also difficult to forget that this profound and desolate piece of music was Sibelius' last masterpiece. After it there were some forty years or so of virtual creative silence, utter bleakness in one way. One could say that the musical structure provides a framework for an intense emotional trajectory involving Sibelius' loss of his active creative life; to use Kramer's take on music and emotion, the music renders tangible Sibelius' bleak emotional drama.

Critics of such an approach to musical meaning may view Ridley's descriptions as mere word painting, however responsive to the music's tone painting, that is free association with little relationship to the true meaning of the work, which is derived purely from its formal musical properties. Indeed, Ridley constantly takes issue with Kivy, who focuses on music's formal properties as accounting for music's effects, and for whom any emotions conveyed by music are at most of what he calls rather strangely the "common or garden" variety, such as anger, fear, and love.[42] For Kivy, there is all the difference between a piece of music being expressive of sadness and a listener actually being sad, a distinction which merely points out that aesthetic emotions consist of a special kind of structured emotional experience, making complex emotional and intellectual situations in life tangible, giving voice to unconscious themes. When one listens to, say, Mahler's *Fourth Symphony*, there is a complex mixture of instrumental, narrative, or programmatic music and song. There are moments of pure contemplation of musical transformations, sudden shifts of mood, continuous tonal movement, sharpness, tenderness, and, in the song, heart-rending resolution, where the various emotional and tonal tensions come together. One is taken along a complex journey of emotion, thought, transformation, and transfiguration, which cannot be encapsulated in terms of simple emotions or mere formal qualities; that would be to reduce the symphony's meaning.

Ridley argues that of course one must take account of music's formal properties when looking at a piece of music's meaning, but that is a mere technical issue, which is helpful up to a point, as appreciation of the technique of voice production or of brush strokes in a painting are indicative of the means by which a singer or an artist makes their own art. But one has to find other words, other vocabulary to get to the heart of the music, as difficult as it may be. In the end, the music had to speak for itself, and in that sense Kivy has a point about accepting music on its own terms, so long as one does not forget all the other aspects of the musical experience, including its emotional trajectory, and also that it involves *performance* between musicians in a setting with listeners, engaging the listeners in what can potentially be a life-changing experience.

One could say that listening to purely instrumental music is rather like looking at an impressive building. You can admire its form, its structure, the way it is put together, as well as the overall frame. This may elicit awe and admiration, or pleasure in seeing a job well done as well as enjoying beautiful proportions and designs. Or is there more to musical appreciation? Is that "more" because we are clearly moved emotionally when words and music go together as in opera and song, so we *expect* the same of music without words? Is it because pure music has some of the dramatic structure of music with words, that we "read" into the music more dramatic structure than it merits? Yet there are clearly specifically musical dramatic elements, such as shifts of tone, volume, consonance and dissonance, and moments of surprise, so is there a specific musical drama, or a mixture of what is unique to pure music and what is borrowed from music with words? There is no easy answer to these questions, but if there were one, it would probably depend in part upon the model of emotion used to account for music's power to evoke or express emotions. Emotions, as I have emphasised, involve complex processes over time, not merely brief discharges of feeling; they often involve some kind of narrative structure, or narrative-like structure, embedding them in human relationships. Unhooking musical listening from this human context and only considering the formal element of the listening process is then artificial and misses out the essential aspect of why we listen to music at all.

When one talks with musicians and composers I doubt if many would confine the emotions created by music to the ordinary garden variety, though one must say anyway that anger, fear, and love are pretty complex emotions and that their expression covers a vast range of possible human scenarios. The French composer Jean-Claude Risset, in a recent collection of papers devoted to the emotional power of music put it well. "I would …

add that while music can arouse more ordinary emotions, most emotions that music arouses in me are quite different from everyday emotions, even though they may have similar bodily effects when they are very strong: weeping or joy ... shivering ... I feel the expressive qualities of music have the capacity to hint at transcendence and to elicit feelings that take us beyond ourselves ... Perhaps music has the virtue to speak to us about what is beyond us, beyond our grasp of our conscious understanding and intelligence."[43]

This resonates with the subtle and dialectical thought of the French philosopher and musicologist Vladimir Jankélévitch for whom, "... music is the domain where ambiguity holds sway ... Music does not express any meaning that can be assigned to it—and nonetheless, music is ... expressive, powerfully so."[44] Music remains elusive in its meaning yet full of meaning, expressive yet also *ineffable*. Music has the power to penetrate to the centre of the soul, as Plato said, and gains possession of the soul in the most energetic fashion. "By means of massive irruptions, music takes up residence in our intimate self and seemingly elects to make its home there."[45] Music for Jankelevitch then enchants us. It expresses infinitely that which cannot be explained, through the magical transaction of the musical performance. Music "... creates a unique state of mind, a state of mind that is ambivalent, and always indefinable. Music is, then, inexpressive not because it expresses nothing but because ... it implies innumerable possibilities of interpretation, because it allows us to choose between them."[46]

For Jankelevitch, the music of Debussy but particularly that of Fauré and Mompou resonate for him with these notions of music's ineffability.

> Debussy evokes movement as a general thing, the pure undetermined essence of mobility, whether the moving object is a breathless runner, or a dead leaf being chased by the wind, or a top spinning in place; the whirling triplets tell of circular motion in the abstract. And this is true as well not only when music is expressing a meaning it claims to suggest to us but also when it expresses emotion, an emotion that it succeeds in inspiring within us. For the expression of feeling in Fauré is as indeterminate as Debussy's description of landscapes ... Fauré calls one of his pieces *Pièces brèves* "Allegress," "Happiness": the C major arpeggios, arpeggios with wings, flying from one end of the white keys to another, express unmotivated joy, indeterminate joy without a cause. In a piece from *Dolly* called

"Tendresse," the two pianists conduct a canonic dialogue without exchanging precise ideas: this could be the soul, silent, conducting a monologue with itself.[47]

One could indeed say that music of all the arts is most able to give shape to the elusive human subject or soul. I have already quoted, in Chapter One, Proust's description of music as the form which is able to capture the communication of souls. The philosopher Roger Scruton points out that he uses the word "soul" on several occasions to try to capture the essence of Mozart's music. Mozart's is not a simple soul; his music "… explores every mood, every character, every turn of the human spirit".[48]

Music for Scruton belongs uniquely to the intentional sphere, a virtual world and not to the material realm,[49] and hence he is deeply sceptical about the value of scientific approaches to musical understanding. As he points out,[50] the basic units of the musical experience are not so easy to grasp. On the one hand, we hear sounds, but music makes use of particular kind of sounds; these sounds are organised by a human subject as a form of communication into an event. Scruton calls this an *"acousmatic event"*, the term being derived from the description in Ancient Greek of the Pythagoreans who were *akousmatikos*, willing to hear. The acousmatic event is to be distinguished from a physical event; it is heard apart from the everyday physical world. In this "other world", sounds are transformed into tones (or pitches); a tone is a sound which exists within a musical "field of force". In other words, the musical world is one where we are in communication as human beings; this is another world from that of the physical world of acoustics. Or as Victor Zuckerkandl put it, as we listen to, say, a violin playing, "As tone follows tone, as the tones become melody, in the midst of the audible world a door opens; we enter, as though in a dream or a fairy tale, not so much into another world as another mode of existence within our familiar world. The audible has broken its ties with material objects."[51]

The composer Thomas Adès captures something about this quality of music in his descriptions of composing. What animates music, he says, is stability and instability. Music is an endless search for stability, trying to fix something that in life would be appreciable only for a moment.

> I can hear a single note and feel all the directions it wants to move in … [E]ssentially the note is alive and therefore unstable … I don't believe at all in the official distinctions between tonal and atonal

music. I think the only way to understand these things is that they
are the result of magnetic forces within the notes, which create a
magnetic tension, an attraction or repulsion. The two notes in an
interval, or any number of chords, have a magnetic relationship of
attraction or repulsion which creates movement in one direction
or another.[52]

Then when it comes to the writing of music, Adès says that there is the
instructive and emotional level but also the analytical level. The latter is
involved when you the sense the form internally and have to find a way to
realise it, which he compares to seeing the face in the fire; the face is not
there, it is virtual, but once written down it becomes real. "Writing music is
like trying to draw the face in the fire."[53]

Michel Imberty[54] defines music as constituting a dynamic system,
which cannot be formulated by cognitive categories because it involves
events and not objects. Of course, one cannot omit the physical reality of
sound. For example, different musical intervals have different patterns of
overtones. A minor third has a remote overtone, which contributes to its
particular quality of disturbance compared to a major third. However, a
major third or major key in one context can feel triumphant but in another
disturbing. It is not the physics that creates the music, but the musical
context occurring in an imagined space. Thus, Simon Rattle describes the
end of Sibelius' *Seventh Symphony* as "almost like a scream ... It's the most
depressed C major in all of musical literature. There's no other piece that
ends in C major where you feel it's the end of the world ... it doesn't sound
like a victory, but as something you reach on the edge of death."[55]

Deryck Cooke has a complex theory of musical expression in his book
The Language of Music, which I think has often been misinterpreted as one
which proposed a simple correspondence between particular emotions
and different musical intervals, such as a minor third always conveying
sadness and a major third a more positive emotion. He is certainly against
the notion that music has no connection to anything other than itself, the
"pure music" theory. On the contrary, he is clear that music conveys human
emotion in many and subtle ways, often, in tonal music, with the use of
an extended lexicon of musical means, a kind of musical language. Thus,
it is true that there are certain formulae that a tonal composer will use to
convey certain kinds of emotional states, such as the C minor theme built
around the minor triad to convey tragedy, as seen in Beethoven's *Eroica*

symphony.[56] He gives many musical examples where specific musical means produce emotional states, but not by a one-to-one correspondence between an interval and an emotion, rather the composer uses the emotive powers inherent in various tonal tensions. Thus, in context, such as with the Sibelius final C major, a major interval can produce pain as well as triumph. Cooke gives the example of the end of Verdi's *Aida*, where the two lovers are about to die walled in their tomb; there they bid farewell to the life of pain and tears. Their violent longing for joy finds painful expression in the initial leap of a major seventh, which is then followed by various rising and falling intervals expressing their intense emotional state. Thus here, "by choosing the right tensions, a composer can make the major key express a considerable intensity of pain".[57] Cooke does add here that the situation in this opera is even more complex as Verdi's lovers are both in pain but also glad to die together, painful though their unsatisfied longing to love may be. The music's tonal tensions convey the mixed emotional states, the mixture of pain and joy of the opera's end, something which music is adept at being able to express.

Thus, the composer "does not express pleasure or pain simply by using the major or minor system, but by bringing forward and emphasizing certain tensions in these systems, in certain ways. This emphasis and these ways derive entirely from the use of the vitalizing agents—volume, time and intervallic tensions."[58]

One can find many examples of how music is used to convey emotions in Cooke's book, but also in *The Treatise on Instrumentation*, originally written by Hector Berlioz but then revised by Richard Strauss. Just to give one example from the latter, Strauss points out how the *tremolo* on the strings can produce a variety of effects. "It expresses unrest, excitement, terror in all nuances of piano, mezzoforte, or fortissimo if it is employed on one or two of the three strings, G, D, A, and if it is not carried much above the middle B flat."[59] Strauss cites this use in the beginning of the monotonously raging storm at the beginning of *Die Walküre*, where the whipping of the rain and hail by the wind is wonderfully depicted in repeated string tremolos. It can also be noted at the beginning of the second act of *Tristan and Isolde*, where the tremolo effect near the bridge of the violin (*am Stege*) depicts the rustling effect of the leaves and the blowing of the wind, producing a "feeling of awe and apprehension in the listener".[60]

Roger Scruton has written extensively and in depth on music from a variety of viewpoints, as a philosopher, musical interpreter, and as an occasional composer, and I cannot possibly do justice to the richness of his thought. But I would extract one unifying theme from his corpus—the issue of music as a virtual or intentional space.

The acousmatic world of music is an *intentional* not a physical space; music has an "aboutness" of its own.

> One of the most important facts about music … is that it is a thing to be *understood*, and understanding music is not a matter of exploring neural pathways or acoustical relations, but a matter of attending to and grasping the intrinsic order and meaning of events in musical space. Furthermore, music is an *appearance*. If you look for music in the order of nature, you will not find it. Of course, you will find sounds, which is to say pitched vibrations, caused as a rule by human activity, and impinging on the ears of those who listen to them. But you won't find any of the features that distinguish *music*. For example, you won't find the space in which music moves. You won't find the gravitational forces that bring melodies to rest or make the notes of a chord cohere as a single entity.[61]

Thus, music is a reality, in which metaphors of space capture its essentially subjective character, and so cannot be grasped from the ordinary cognitive standpoint. "No science, no theory, no form of explanation with which we order and predict the physical world, could possibly make contact with the movement we hear when we hear a melody in musical space."[62] Just as the painter who applies pigments "… in one space creates a world of imaginary beings in another, so does the musician [who makes] sounds in our space create virtual actions and movements in the acousmatic space of music".[63] This is the reason we feel, as Mahler described, that music puts us in touch with another world. This is also perhaps why we may feel that music reaches deep into our soul, or what I have described as our psychic home.

One can see in Scruton's approach to the nature of music the profound influence of the idealistic aesthetic thought of the Italian philosopher Benedetto Croce. In his *Guide to Aesthetics* Croce describes art positively as *vision* or *intuition*. "The artist produces an image or picture. The person who enjoys art turns his eyes in the direction which the artist has pointed out to him … and reproduces in himself the artist's image. 'Intuition', 'vision', 'contemplation', 'imagination', 'fancy', 'invention', 'representation', and so forth, are words which continually appear as almost synonymous in discussions on art. All of them give rise in our minds to the same concept or to the same set of concepts—a sign of universal consent."[64] Art is also distinguished from material reality; it is not a physical fact because for Croce such facts lack reality. "On the other hand, art, to which so many devote their whole lives and which fills everyone with heavenly joy, is *supremely real*. Consequently, it cannot be a physical fact, which is something unreal."[65] Similarly for Scruton music is not a physical fact, even if it consists of physical sounds; the latter do not make music. The acousmatic event in which the human subject is immersed is to be distinguished from a physical event; it is heard apart from the everyday physical world.

For Croce what gives coherence and unity to the artistic intuition is *intense feeling*, coupled in some way to a representation. "Intuition is truly such because it expresses an intense feeling, and can arise only when the latter is its source and basis. Not idea but intense feeling is what confers upon art the ethereal lightness of the symbol. Art is precisely a yearning kept within the bounds of a representation."[66] Expression and intuition are one and the same. "In reality, we do not know anything but intuitions that are expressed. A thought is not for us thought unless it is formulated into words. Neither is a musical fancy unless it is made concrete through sounds, nor a pictorial imagination unless it is rendered in colour."[67] The expression of a living work of art cannot be detached from the form in which it is expressed; every work of art has its own particular law and its full and irreplaceable value. As Scruton puts it when discussing Croce's thought, "Expression must be grasped in the *particular* experience of the *particular* work ... In which case, the only way to identify *what* is expressed by the last movement of the 'Jupiter' symphony, is to play the last movement of the 'Jupiter' symphony."[68]

The *intensity* of the artistic vision then is what is responsible for its power; that intense vision invites the viewer or the listener into the orbit of the work, engaging us to respond to the particular vision in an essentially intersubjective relationship between the work and the observer or listener in a special imaginative space.

Scruton's own take on this essentially Crocean model, albeit with modifications, is summarised in his recent book *Music as an Art*, when discussing the idealism and philosophy of music.

> ... sounds become music when they are organized in such a way as
> to invite acousmatic listening. Music is then heard to *address* the

listener, I to you, and the listener responds with the overreaching attitudes that are the norm in interpersonal relations. These attitudes reach for the subjective horizon, the edge behind the musical object. The music invites the listener to adopt its own subjective viewpoint, through a kind of empathy that shows the world from a perspective that is no one's and therefore everyone's. All this is true of music in part because it is an abstract, non-representational art, in part because it avails itself of temporal organization in a non-physical space.[69]

Elsewhere,[70] I have described a similar process when contemplating a late Rembrandt self-portrait, where his eyes seem to take you into the picture, into the virtual depths, where you are in touch with Rembrandt's intense and powerful artistic vision. Unlike a mirror, which reflects your own image back to you, the Rembrandt urges you to reflect into yourself in the act of being drawn into his image. Repeated visits are like drawing from some primal source of light and intensity, leaving you changed in some way, both uplifted and more melancholy. There is of course as well the presence of Rembrandt's own eyes, not only those you see in his self-portrait looking out at the spectator, or rather beyond the spectator to some other region, but also those eyes of his which look inwardly so poignantly at himself, scrutinising and accepting what he saw with so few illusions.

The effect of such viewing remains, to me at least, something of a mystery. How is it that the portrait of a dead artist can have such life? How can marks of paint, however cleverly applied, still speak to us over and over, continuously drawing us both into the picture and into ourselves? It is as if we are witnessing some source of inner light in the picture itself. What is that elusive something that makes this happen? What is it in ourselves that is drawn out by repeatedly viewing the self-portrait? Similarly, what is it about profound music that continuously speaks to us and moves us? Is it surely not merely the physical effects of music, its frisson, its tremolo effects, or its stirring of the basal ganglia.

It does seem that we are here in the area of the human "*soul*". It is what the psychologist Nicholas Humphrey has recently called the "soul niche", a "place where the magical interiority of human minds makes itself felt on every side".[71] Though very much rooted in cognitive science, he quotes with approval the theologian Keith Ward from the latter's book *In Defence of the Soul*. Ward makes the case that the whole point of talking of the soul is to remind us that we transcend the conditions of our material existence; we are not just molecules and genes. Thus:

> To believe in the soul is to believe that man is not just an object to be studied, experimented upon and scientifically defined and analysed, manipulated and controlled. It is to believe that man is essentially a subject, a centre of consciousness and reason, who transcends all objective analysis, who is always more than can be scientifically defined, predicted or controlled. In his essential subjectivity, man is a subject who has the capacity to be free and responsible—to be guided by moral claims, to determine his own nature by his response to these claims.[72]

Such a view about what makes us human resonates with the psychoanalytic view of the human subject, as I have explored, along with the soul concept, at length in my book *The Psychic Home*.[73] In brief, Freud's discoveries were very much about bringing back into the realm of the human subject elements of the mind such as dreams and fantasies which had been devalued as mere objects of, at best, some objective knowledge, or, at worst, of no consequence, just debris of the mind. The psychoanalytic encounter is very much about helping the patient "*become a subject*", through a process of recovery, or discovery, of their unconscious subjective life; dreams and fantasies, for example, are precious signposts towards capturing the elusive human subject. I mean by this that the analytic patient brings to the analyst all sorts of different stories, fixed patterns of relating or symptoms, hopes, expectations, and resistances. Patients often come with a sense of isolation; of either being alone with suffering or suffering from being alone. And they come to analysis subject *to* various forces in their life, past, and present. If the analysis works, then there is the possibility of their becoming the subject *of* their experiences and ultimately their lives, with a sense of no longer feeling isolated, while being more in contact with others. Some patients have described this shift as finding themselves, or finding meaning, of feeling real, or of having a centre where before there was chaos or nothingness, and even occasionally that they have found their "soul", or that the "soul" has been put back into their lives, which until then had become "soul-less".

Everyday language occasionally uses the soul concept in somewhat similar ways in order to express something alive in the human subject. Music can be described as having soul when it hits the emotional core of the listener. And, of course, there is "soul music", whose basic rhythms reach deep into the body to create a powerful feeling of aliveness. Once more, one can truly say that music of all the arts is most able to give shape to the elusive human subject or soul.

In *The Interpretation of Dreams*, Freud came up with an image of the mental apparatus which highlights the virtuality of subjective space, albeit using a scientific metaphor. In order to capture the way that the mental apparatus functioned, he disregarded the notion of any anatomical locality; instead, he pictured the apparatus to be like a compound microscope or a photographic instrument. "On that basis, psychical locality will correspond to a point inside the apparatus at which one of the preliminary stages of an image comes into being. In the microscope and telescope, as we know, these occur in part at ideal points, regions in which no tangible component of the apparatus is situated."[74]

Reik described how in psychoanalytic work analysts need to "hear themselves", that is, use their unconscious as a receiving instrument. One way of doing this is to turn to the auxiliary sense of perception experienced in the emergence of musical themes in order to tap into the wires of unconscious life, as, "The intangible that is invisible as well as untouchable can still be audible. It can announce its presence and effect in tunes, faintly heard inside you."[75]

One can also compare the role of the virtual space where music is heard, or announced, to the place of Winnicott's transitional area and virtual space in early mother–child interactions and the place of the transitional area in culture and day-to-day living. In Chapter One, I mentioned how Lombardi described music as connected with both the concrete world of bodily sensations and the symbolic expression of culture, and may then be an important transitional phenomenon that can keep the internal world and external worlds connected. This can be seen in communication between patient and analyst on both conscious and unconscious levels. Winnicott's notion of transitional experience stems from his observation of the important role of transitional objects, such as a blanket or a soft toy, in young children's development. With his notion of a transitional object, the infant's first "not-me" possession, Winnicott described a form of object relationship where the object is experienced simultaneously as created by the infant and as discovered by them. The transitional object comes from outside, from our or the parent's viewpoint, but not from that of the infant. Nor does it come from within; it is not an internal object. The essential feature of the transitional object and transitional phenomena, "is the *paradox*, and *the acceptance of the paradox*: the baby creates the object, but the object was there waiting to be created".[76]

Furthermore, transitional objects and phenomena belong to the realm of *illusion*, which is the basis of the initiation of experience. The adaptive mother allows the infant the illusion that what the infant creates really exists. Thus, a new area of experience, the transitional area, is brought into being by the meeting, or coming together, of two different viewpoints focused upon the baby's needs. The transitional object never "goes inside" nor is it forgotten or mourned; it just gradually loses meaning because "the transitional phenomena have become diffused, have become spread out over the whole intermediate territory between 'inner psychic reality' and the 'external world as perceived by two persons in common,' that is to say, over the whole cultural field".[77]

For Winnicott, we live life in the day to day very much in this intermediate area of experience.

> What, for instance are we doing when we listen to a Beethoven symphony or making a pilgrimage to a picture gallery or reading *Troilus and Cressida* in bed, or playing tennis? What is a child doing when sitting on the floor playing with toys under the aegis of the mother? What is a group of teenagers doing participating in a pop session? It is not only: what are we doing? The question also needs to be posed: where are we (if anywhere at all)? We have used the concepts of inner and outer, and we want a third concept. Where are we when we are doing what in fact we do a great deal of the time, namely enjoying ourselves?[78]

For Winnicott, the answer is that in health we live in the intermediate zone, the third area of transitional space; this is then the virtual space in which music moves and which becomes tangible in a performance. As Mark Wigglesworth described, the "core of a creative musical experience hovers somewhere in the middle of a triangle formed by composer, conductor and orchestra. In some mysterious way, everyone travels towards this musical centroid and it is by making this journey that the uniqueness of any performance is created".[79]

This is also what Marion Milner has referred to as the area of "creative illusion", when one does not have to decide which is, or belongs to, oneself or belongs to the other, but somewhere between both.[80] These moments of creative illusion are integral to the aesthetic moment, for example when I described the contemplation of the Rembrandt self-portrait, when there was a feeling of being drawn into the picture and becoming one with it, while also remaining observing from the outside; inside and outside merge, the boundary between self and other dissolves while also remaining.

Art also provides a method in adult life for reproducing states of mind that are part of everyday experience in young children, the infant with their transitional object or the older child absorbed in play. Because music has its essential roots in preverbal experience it is linked closely to vitality affects. It does have connections to the developing language organisation which does start forming at a young age, as revealed in the way that the baby's language areas of the brain are already active when listening to sounds, but it is not yet organised in a way that will dominate the child's communications. Music is then not at this early stage in the full world of representation but rather in that of the imagination or the "imaginary". The transitional area is still rooted in the imaginary area, and dominated by the primary process, where symbols are not yet fully formed. Later on, when music is organised by the adult mind, such as when it is written down in a score or performed with complex rhythms and harmonies and arranged into melodies, the secondary process is dominant, even if the primary process can still be tapped into; the secondary process controls or channels the primary process, or tries to.

Music, with its close ties to the body, primordial feelings, rhythms, and movement, is most able to encapsulate, or shape, the inarticulate, to communicate what cannot be communicated, those moments of creative illusion, often pleasurable, where subject and other merge. So much of musical enjoyment comes from surrendering to the music, not just because the sounds themselves elicit pleasure, not just because of the effect of the specific emotional content and the play of musical tensions, but because we can lose ourselves in the music while also remaining ourselves; these are the moments of pleasurable illusion. Music has the power in performance to touch the intermediate area, giving us the illusion of wholeness, however briefly, where subject and other merge.

The place of illusion is more clearly visible in theatre, and thereby also in opera. The special setting, the fact that the audience is usually seated and relaxed, that there is usually darkness which predisposes to the formation of a state akin to dreaming, and that the stage is especially lit, create the right atmosphere for illusion to work. One could say that there is a kind of mesmerising transference created by this setting between the audience and the players. The actors, or the singers, are both themselves and the role that they are playing; their dual role is part of the illusion.

The audience is at the receiving end of the illusion process; the performers, as I have already described, are not necessarily in that state of mind, concerned as they are with the actual process of performance; though it is notable that musicians, like all performers, describe moments when they lose all sense of time passing, when they cannot even remember what they did when they produced something exceptional. This may be when the technical aspects of the music are so completely mastered that the performer can trust in their unconscious to take them where they want to go, rather than needing to be in clear conscious control of themselves. But for the most part the performer relies on a constant flow between conscious control and unconscious letting go.

Tia DeNora, as I have already mentioned, has made detailed empirical studies of music's *social* function, for example through detailed question-naires given to a variety of people. She describes how "recent studies of music in relation to the achievement of emotional states and events point to music's use, in real social settings, as a device that actors employ for entraining and structuring feeling trajectories. Music is a resource to which agents turn so as to regulate themselves as aesthetic agents, as feeling, thinking, and acting beings in their day-to-day lives."[81] Music thus "channels" emotions in a variety of ways and in a variety of social settings, whether that be to manage the stress and strain of the daily commute, to enhance communal experiences in religious ceremonies, to help people get "in the mood" for an event, or to influence consumer choice in supermarkets. Music has a significant role in the maintenance of mood, memory, and even identity, for example through musical entrainment, synchronising body and mind.[82]

She cites a number of features of music which make it particularly suitable for evoking an emotional response,[83] and which also gives music its significant social power. These are first its *materiality*, its physical properties, which often involve human bodily action and reaction; music has direct effects on the body—I have also described the direct effect of music at a brain level, such as the brainstem and navigational system. Then there is music's *iconicity*, that is, that it often shares structural features with things it is seen to represent, as can be seen in, for example, the way that John Dowland's lachrymosal pieces flow, melodically, downward, like the tears they describe. Next there is the existence of tradition and *convention*, where certain kinds of music come to represent certain kinds of response, such as with church music. Then, there is the fact that music moves in time, that its *temporality* means that music's power in relation to emotion derives in part from its ability to portray psychological movements or subjective processes. Other aspects of music's ability to convey emotion include the role of *expectancy*, as I have already covered in Chapter Four, and then its ability to convey ambiguity or to be effectively *non-representational* yet still convey meaning. In sum, it is part of music's musical character that can make it suitable as a "device for eliciting, evoking, channelling, representing, and/or inducing emotional response".[84]

Jenefer Robinson takes as central to her theory of artistic and musical expression the idea that expression is a kind of elucidation and articulation of emotion, and suggests that the cognitive process of becoming clear about an emotion is part of the actions of musical listening and performance. She argues that music like emotion is a *process*,[85] and so it is particularly well suited to express not only particular emotional states but also different kinds of emotion, conflicts between emotions, ambiguous emotions, and the way that one emotion may transform into another emotion. If emotions are processes and our emotional life occurs in streams, then she argues that it is reasonable to think that music, which is itself a process or series of processes occurring in streams, is particularly well suited to mirror emotions, for example through complex movements of harmony, rhythm, and melody.

When we have an emotion, she describes a process where there is an immediate affective "appraisal", which can be fast and automatic, be experienced at a body level, and for which one can find clear physiological evidence. This is followed by a slower cognitive evaluation, which may be or become quite sophisticated. The emotion involves the whole process, and thus emotions are comprised of affective and cognitive elements, not one or the other, or one element at the expense of the other, similar to the interactions between primary and secondary process in the mind. Thus, when we listen to music, we have the immediate affective responses and also the cognitive appraisal where we can reflect upon the responses; we can come to grasp the structure of the music as well as what it expresses.[86] We feel the music but we also judge it in some way. This is a kind of active listening, allowing the music to interpenetrate one's being while also reflecting on what we experience. As Aaron Copland[87] wrote, "In a sense, the ideal listener is both inside and outside the music at the same time, judging it and enjoying it, wishing it would go one way and watching it go another ... A subjective and objective attitude is implied in both creating and listening to music." This is very near to analytic listening, where we allow the experience of being with the patient to have an impact on us at various levels.

The final chapter will focus on connections between music and psychoanalysis as a way of drawing together the various threads of the book's themes.

CHAPTER SIX

Finale—musical
and psychoanalytic connections

Musical creativity

A common anxiety or suspicion about psychoanalytic treatment is that it may spoil creativity, that an artist or musician may lose something vital if they submit themselves to psychoanalytic help, and that psychoanalysis is too reductive to be of help to those who need to find new ways of seeing the world. I certainly agree that taking a creative person into analysis brings with it an extra responsibility to respect areas of functioning that have worked in the past, but the reality is often that creative people seek help when they are blocked or inhibited creatively, or, with actors and musicians, when they have anxieties about their performance. Julie Nagel has written specifically about the treatment of stage fright in musicians. She describes how a confluence of narcissistic sensitivities, ego/superego development, shame dynamics, physical injury and pain, and terrors pertaining to psychic and body disintegration can fuel anxiety reactions in many people, but that these dynamics hold particular resonance for the performing musician who has spent formative years preparing for a musical career.[1]

My focus in the book, however, has been, with a few exceptions, less on the direct clinical relevance of psychoanalysis to musicians than to the more general issues about the connections between music and psychoanalysis. There are now a number of books and papers tackling the general relationship between music and psychoanalysis, some of which I have already cited. After a number of years with little written specifically about music, two collections of psychoanalytic explorations came out in 1990 and 1993,[2] covering psychoanalytic contributions to the psychology of music, applying analytic theory to musical composition, the relationship between music and affect, and a large number of studies of composers and compositions. Since then, there have been a few scattered papers concerning music but also some substantial books on the general relationship between music and psychoanalysis, such as those of Antonio Di Benedetto, Gilbert Rose, Michel Imberty, and Julie Nagel.[3] I have already referred to Theodor Reik's early work showing, with many clinical examples, how musical associations arising in the analyst's mind can be of great help in the understanding of the patient's communications. But the first to write about music and psychoanalysis was Max Graf, an author and musicologist, the father of "Little Hans", the five-year-old with a phobia of horses and the subject of Freud's classic case history, "Analysis of a Phobia in a Five-Year-Old Boy". Freud helped Hans to overcome his phobia indirectly, by means of various suggestions conveyed to his father. In fact, Little Hans, or Herbert Graf, whose godfather was Gustav Mahler, became a successful opera producer and produced operas at the New York Met for many years and in various other countries, including Covent Garden in London. His father was a distinguished musical historian and critic, and played an important part in Vienna's rich musical life in the late nineteenth and early twentieth centuries, until he moved with his family to the US following the rise of the Nazis.

Max Graf was part of Freud's early circle, and attended the Wednesday night scientific meetings and presented there a number of papers on music and psychoanalysis.[4] Much of his thought is encapsulated in a book he subsequently wrote while in the US—*From Beethoven to Shostakovich (The Psychology of the Composing Process)*.[5] It is a book rich in musical examples from the Western art canon, and a deep knowledge not only of classical composers such as Haydn, Mozart, and Beethoven, but also of late romantic composers, a number of whom he knew personally. Graf's aim from the beginning of his association with Freud was to take over the task of investigating through psychoanalytic understanding the psychology of great musicians and the process of composing music. I have in the main avoided directly applying psychoanalytic understanding to the life of composers, except in the rare instance, such as with Mahler, where we have considerable personal information and where scholarship has already delved deep into their lives. Nonetheless Graf's position as one central to Vienna's musical life did put him in a unique position to make convincing points about the nature of musical creativity and composition, still relevant today.

Graf finds the sources of musical imagination in the depths of man's emotional life in the unconscious, the deepest layer of man's soul. Thanks to the composition and performance of music, emotions "usually lying chained in the dark of the unconscious break loose from their prison".[6] Rhythm puts order in the process of musical fantasy, it imparts to music life and unity but also the power to "arrest the audience, to keep them tense, to hold them in an exalted mood".[7] For Graf, music is the most powerful means of producing an intensification of the emotions, and of expanding and enhancing them, and of "directing the magnified tensions of the soul toward one goal".[8]

Graf attempts to describe the composer's state of mind, their organising power of personality that can produce musical works of art, delineating how the composer reaches down into their unconscious, tapping into childhood memories and experiences and external influences to produce their most important creative work. He suggests that there are three overlapping phases in the composer's work—the preliminary work done by the unconscious from the often chaotic sentiments and passions that are to become melody and tone, as well as elements of personal life from childhood onwards; the combined work of unconscious and conscious mental powers; and the conscious final polishing of the form.[9]

> The artist possesses the strength of organizing with his imagination the impulses of the unconscious that attack the borders of the conscious ... He transforms them to shapes, patterns, melodies ... That which in mental illness lacerates the soul, or forcibly invades it, or makes it a toy of uncontrolled emotions; and that which in dreams joins one seemingly meaningless picture to another, in the work of art becomes intelligently organized form and logical unity that connect fantasy and thinking.[10]

To create musically means not "to be driven by moods, but to be able to concentrate moods and summarize them, be it in an ideal picture as those created by musical classicists, or in a fantastic picture as done by the romanticists, or in a realistic portrait such as Richard Strauss would paint, and in the case of Claude Debussy, with impressionistic technique".[11]

Graf uses the notion of a "psychic complex" that was prominent in psychology early in the twentieth century; this refers to a set of memories and emotions that cohere together in some way and help to shape emotional life, often around a unifying theme. One can see the "death complex" having a pivotal role in Mahler's music, counterbalanced by his intense vision of nature and life overcoming anxiety over death. For example, there are a number of funeral marches in his music, the first movement of his second symphony opens with a "Dies irae" ("the Day of Wrath"), he even, as I mentioned in Chapter One, composed the *Kindertotenlieder* ("Songs on the Death of Children"), and in the last movement of the *Fourth Symphony*, where, using one of the *Wunderhorn* poems, a child (soprano) sings of a blissful life in heaven. At the close of the fourth movement of Mahler's unfinished *Tenth Symphony*, "a muffled drumbeat returns: it is the rolling of a drum in a funeral procession that passed Mahler's New York hotel and the beat of it moved Mahler to tears".[12] It is possible that Mahler's preoccupation with death was linked to the early death of six of his siblings, as well as the experience of having an abusive father.

Graf suggests that Wagner has a repeated constellation of themes organised around the desire for "salvation through love", as his way of sublimating his strong libidinal drive. He first came across this in the motif of the *The Flying Dutchman* redeemed by the love of a woman in the works of the poet Heinrich Heine. It is this longing, repeated time and again that "creates the tension in the music of 'Tristan and Isolde' and finds its release in Isolde's 'Liebestod.' It inspired Wagner for the ending of 'Götterdammerung,' where the world is redeemed of the curse as Brunhilde leaps into the flames. It is the chief motif of Wagner's poetic creation."[13] He points out that the operas, *The Flying Dutchman*, *Tristan and Isolde*, and *Götterdammerung* all close with the same harmonies—a minor subdominant and suspension over the basic triad, "musical expression of yearning that contracts painfully once more before dying away in spiritual peace".[14]

In speaking of great artists, Graf considers that one can distinguish between two kinds of experience—"elemental experiences", which come from the recesses of their soul and grow out of the roots of their personality, including childhood experiences and traumas, and "formative experiences", which originate in external life. But the ability of organising all these experiences into a great work of art ultimately defies scientific analysis.

This does not prevent Graf from suggesting how composers go about their composing. Artistic work sets in once the work of the unconscious has begun to release or generate various musical thoughts, some already formed, others less so. The composer may need to fall into a "productive mood" in order to facilitate the opening of the usually well-guarded door leading from the unconscious to consciousness, though that might take time, with a great deal of unconscious work already generating ideas deep in the composer's mind. He quotes Richard Strauss as describing how "Musical ideas, like young wine, should be put in storage and taken up again only after they have been allowed to ferment and ripen. I often jot down a motif or a melody and then tuck it away for a year. *Then when I take it up again I find that quite unconsciously something inside me—the imagination—has been at work on it.*"[15]

Graf calls "musical conception" the moment in which elemental unconscious musical figures break through and are seized by the composer's consciousness; from then on conscious thought and critical thinking come into play, leading to the final work. He illustrates this process with musical examples, such as when Beethoven reworked the overtures to his opera *Fidelio*, moving from different key choices, until one key, E major, suddenly emerged from his unconscious to capture the sense of coming into the light from the darkness, a major theme of the opera. Thanks to the preservation of Beethoven's sketchbooks, we do have evidence of how certain harmonic modulations and motifs formed the basis for subsequent formed works, along the lines that Strauss describes with his own creative process. One can see how different motifs, keys, modulations, and melodies are tried out, suddenly appear, and are revised or abandoned, as the music begins to take shape.

Graf also outlines the nature of the final stages of composing, when artistic reflection on the musical material takes place, involving reflection before the actual creative work begins, such as shown in Beethoven's sketchbooks, reflection accompanying the work, with revisions and adaptations, and then the final acts of reflection when the finishing touches are made to present the completed work. Actual composition is "accomplished in a regulated coordination of unconscious forming and critical thinking, of inspiration and work. This harmony of the creative forces is the most difficult of compositional work. It may be disturbed at any moment, and requires an uninterrupted balance of the conscious and unconscious faculties."[16]

Finally, he suggests that the work of the composer, which fills the work of their imagination more and more with the light of consciousness, can be embraced in various individual activities—condensing and simplifying, expanding of ideas and forms, and elaboration and intensification of every detail, all put together with the aid of the composer's technique. This last capability is partly learned and partly inherited, and moulded by their own personal force, or their artistic sensibility, the life of their soul."[17] One could add that the composer would need to have a finely tuned "ear" for music, a feeling for sound and its effects, as well as an emotional rapport with the listening audience.

One psychoanalytic concept particularly relevant to the composing process, and indeed to other forms of artistic endeavour, as a way of describing an essential element of how unconscious thoughts and feelings are worked over and put to artistic use, is that of "*working-through*," or *Durcharbeiten* in German. Much of what Max Graf wrote about the composing process basically illustrates working-through, with unconscious thoughts being released and then worked on by the composer's conscious processes.

Working-through, as René Roussillon writes,[18] is an integral part of the psychoanalytic process, even its epitome. It was a term used by Freud to describe the process by which the work of analysis, through interpretation, overcomes the patient's resistances to accepting their unconscious thoughts and feelings. The idea that analysis involves psychic "work" was an early one in Freud. Thus, in *Studies on Hysteria* he describes how in the hysterical patient there is a nucleus consisting of traumatic memories of events or trains of thought, and around this nucleus is an agglomeration of memories which have to be worked through in the analysis in order to relieve the patient of their hysterical symptoms.[19] *The Interpretation of Dreams* has a whole extended section called the "Dream Work",[20] which refers to how latent dream thoughts are transformed into the manifest dream—through the work of condensation, displacement, considerations of representability or selection, and transformation into visual images, and secondary revision which puts the dream into some sort of order.

It was only in his 1914 paper, "Remembering, Repeating and Working-Through", that working-though became a fundamental element of treatment, when he describes the expenditure of work that patients have to make in order to overcome their criticism of their free associations, or their "resistances".[21] After this process, past memories, for example of traumatic situations, may be remembered rather than continuing to be repeated in an unmodified form. The working-through of these resistances can be an arduous task for analyst and patient, but it is nevertheless a "part of the work which effects the greatest changes in the patient and which distinguishes analytic treatment from any kind of treatment by suggestion".[22] After the formulation of his second topography, Freud subsequently defined different kinds of resistance arising from different parts of the mind, from the ego, id, and superego but the basic notion of working-through of unconscious resistances remained essential to analytic treatment. It points to a process whereby unconscious processes and formations are transformed into a form which is amenable to the human subject's consciousness; they can become subject "of" their experiences.

Subsequent analysts have emphasised that working-through is not just a part of the patient's work but also that of the analyst; that, for example, working-through in the analyst's countertransference is a vital part of the analytic treatment; both patient and analyst have to work though their unconscious resistance and evasions. For example, Irma Brenman Pick illustrates how analysts need to allow themselves to have various disturbing experiences evoked by the patient and then digest them, and then formulate this as an interpretation—all involving psychic work.[23]

Thus working-through involves some form of psychic transformation of unconscious material into a form more acceptable to consciousness, involving both patient and analyst. Roussillon suggests that this work of processing is similar to the importance of play for children, as described by Winnicott, "because it has the same function as play at that stage of life: to bring under control difficult and potentially traumatic situations in order to symbolize them and prepare the way for their subjective appropriation",[24] or what the French call "*subjectivication*", or the owning of one's own psychic reality, or what I have called "becoming a subject". The aim of French psycho-analysis tends to be that of freeing up the associative process, facilitating free association, which then allows working-through to take place.[25] This would seem to resonate with how, at least prior to the detailed elaboration of a musical score, a composer moves from being receptive to vague thoughts often arising from the unconscious, which then release or generate various musical thoughts, some already formed, others less so, followed by the hard work of processing and elaborating the material into an artistic form.

Darian Leader[26] has shown how working-through has particular musical resonances, and how a sensibility towards musical concepts can help to formulate clinical issues. One can see a kind of working out or working-through taking place in the way that music is structured by sonata form. After the introduction of the principal themes, there usually follows the development section, which treats the themes to elaboration. The themes may be disintegrated into constituent parts and then combined and varied before returning in the final cadence to the initial theme. "Although the conclusion, or recapitulation, restates the initial themes, it may include changes arrived at during the development. There is thus a real tension between working through and repetition, a tension that musicologists have sometimes situated as the key dynamic of the sonata form and which we find in the title of Freud's [1914] paper."[27] Like a musical score, working-through in analysis involves the return of what was there at the start, but it has transformed the material into a new configuration.

Martin Nass has made a study of a number of American composers and their creative processes, interviewing them in order to try to capture the inspirational phase of their work. He makes the point that the best manner of studying the creative process is to speak with the creators themselves, rather than, say, trying to link creative activity with illness or pathology,[28] which a number of psychoanalysts and other psychologists have tended to do, though not one must add Max Graf, who of course knew personally, or was in contact with, a number of the significant musicians of his time.

Nass found the presence of a number of key elements in the development of the creative imagination of these composers—heightened sensory awareness, together with a particularly acute capacity to retain non-verbal childhood experiences; the capacity to deal with loss and to work through issues of mourning; to be able to take a difficult path, even when a less anxiety-provoking one was available, and to own their work rather than feel passive—or what I have emphasised as being a subject *of* their experience; and the capacity to tolerate and maintain ambiguity and a lack of closure.[29]

Nass found that there was no single way that gifted composers write, that they aren't always auditory people but tend to use a variety of sensory skills, and that there are as many individual variations in working practices as there are differences in personality, but that what appears common is the ability to tolerate uncertainty and doubt, to take chances, and to be able to wait until something arises, that is, what Graf had described about their being able to allow a door to be opened from the unconscious to the conscious mind, something, one might add, that is a key element of psycho-analytic free association within an analytic session.

The American composer Roger Sessions, one of those in fact whom Nass had interviewed, wrote about the musical experience of the composer. For him, the task of every composer is to "give coherent shape to his musical ideas: or as Artur Schnabel has so finely put it: 'The process of artistic creation is always the same—from *inwardness to lucidity.*'"[30]

Sessions describes the "musical idea" as simply "the fragment of music which forms the composer's point of departure, either for a whole composition or for an episode or even a single aspect of a composition".[31] This musical idea, a fragment or more complex musical element, starts off the musical train of thought. Composers, and a number of musicians, often have such thoughts moving in their head. One could see here a similarity with psychoanalytic free association, where one thought leads to another and in that movement of thought unconscious meanings, often linked to powerful emotions, emerge. Indeed, Sessions describes the beginning of a musical idea as a "train of thought",[32] its only difference from any other train of thought being its medium being tones, not words or images. One can see this clearly in Beethoven's sketchbooks, where musical motifs are progressively shaped into larger structures. For Sessions, the musical train of thought is largely a "train of impulse or feeling" and is also a complex mixture of conscious and unconscious processes. Though this is also true of many non-musical trains of thought where ideas and images are highly charged with emotion and that emotion frequently motivates the sequence of ideas, in music it is the logic of sensation and impulse that determines the ultimate validity of the train of thought and gives the musical work not only its expressive power but whatever really organic unity it may possess.

He also proposes some underlying principles of the composition process.[33] First, there is a principle of progression or cumulation, where, music being an art of time, its effect must be cumulative and not static, otherwise it will sound boring. Second, is the principle of association, including the use of repetition. A musical motif only takes on meaning through association with other motifs or phrases; it means nothing in itself. This includes both association with other musical elements but also with words. The principle of association gives significance to musical ideas and unity to musical forms. Lastly, the principle of contrast throws musical ideas into relief, usually involving only one major dominating contrast in a work of movement, where basic forces are brought into opposition, presumably leading to some sort of resolution. The composer brings their technique into play in order to create the final composition, which consist of their basic craft, their learned know-how and mastery of the musical language, but in addition they will become not merely a craftsman, but a musical thinker, a creator of important human values.

Aaron Copland, in discussing the creative process in music, notes that every composer begins with a musical idea, for example a theme or melody or series of rhythms, which may suddenly come to them. Not only does the composer then examine the musical line for its formal qualities, such as its beauty, the way it rises and falls, but also for its emotional significance.[34] For him, all music has expressive power, though precisely what music expresses cannot be put precisely into words. "Music can express at different moments serenity or exuberance, regret or triumph, fury or delight. It expresses each of these moods, and many others, in a numberless variety of subtle shadings and differences. It may even express a state of meaning for which there exists no adequate word in any language."[35]

If all music has expressive value, then the composer, through conscious thought or through feeling, gets to sense the emotional nature of his theme, and how changing the dynamics or harmony or rhythm can alter its emotional tone and significance, and how capable it is of being transformed in interesting and significant ways.

There are those composers, he suggests,[36] who begin not so much with a musical theme but with a completed composition, particularly if they are composing a shorter piece such as a song, as can be seen with Schubert. There are also those such as Beethoven, but also the majority of composers, who are more of a "constructive type", who work over themes and gradually build up a completed work. And there are the so-called "traditionalists" who begin with a pattern rather than a theme, such as Palestrina, for whom the creative act was not the thematic conception so much as the personal treatment of a well-established pattern, though he also found ways of adapting the pattern to new circumstances, for example when he found a way of rescuing polyphony from the structures of the post-Reformation Catholic Church. The Church had demanded the clear expression of words in sung music so that the faithful would be in no doubt about their meaning; polyphony had become so complex that the Church feared that the words' meanings were being distorted, at a time when the Catholic faith was being challenged by Lutheran Protestantism and its emphasis on simplicity of worship. Palestrina rescued polyphony by setting the music to the words' syllables, producing both beautiful and complex music and intelligible words. Finally, there are those who are pioneers, such as Gesualdo, Debussy, and Varèse, who oppose conventional solutions to musical problems, seeking new harmonies, sonorities, and formal principles.

To get from the musical idea to the composition requires a process of "bridging", such as elongating a theme, and then welding together all the material into a finished product. For the latter to work effectively, Copland refers to what he was taught as a student, perhaps by Nadia Boulanger when he studied with her in Paris, which is to look for *la grande ligne*, the long line. This is something the composer has to *feel*. It refers to a sense of flow, a sense of continuity from first to last. "Music must always flow, for that is part of its very essence, but the creation of that continuity and flow—that long line—constitutes the be-all and end-all of every composer's existence."[37]

One can see various kinds of musical line in different musical eras. For example, with opera, the musical line in Wagner tends to flow continuously. By Puccini, that line can suddenly shift as the moods and the characterisation shifts, making for moments of sudden excitement. While as the twentieth century progressed, the musical line can become increasingly fragmented as complex and conflictual states of mind are represented, such as in Alban Berg's opera *Wozzeck*.

Searching for the musical line, the meaningful connection between pitches, is also an essential part of the work of the performer, whether that be a singer or an instrumentalist, or a conductor. That also entails forming a bond with the music, trying to connect with the composer's vision.

With regard to the relationship between the composer and their personality or personal experiences, I have already suggested that it is not easy to think about the place of the composer as a subject and their relation to their music and that one must be cautious about drawing inferences concerning the relations between a composer's life and their music, unless there is clear documentary evidence of links. There are some composers whose work and life seem inextricably linked, but with others, the life and the art may seem to be only partially connected; the music may function as a separate realm of creative endeavour or illusion, protected from the turmoil or muddle of daily life.

I have also, in Chapter One, cited Edward Cone,[38] who suggested that the composer makes use of a "persona", or assumes a special role, in order to communicate. While persona and composer as person are linked, the persona also represents what one could see as a special state of mind, or a projection of the composer's musical intelligence, a kind of self-observing ego used in acts of composition. We may never know, or need to know, the actual person of the composer. Yet we do know the composer's persona, what makes their music recognisable, their style or signature, their voice. In the performance of a vocal piece of music there are a number of different personas or voices—as we have the voice of the composer behind the music, the voice in the music, for example a character, and the actual voice of the performer who themselves will use a special frame of mind to communicate the meaning and emotions of the music, which may or may not be directly related to the emotions of the song.

Music and emotion

Some psychoanalysts have tackled how music conveys emotions. As Julie Nagel describes, music can have deep and transformative effects, not only in loosening defences but also, as she shows with clinical examples, in deepening life and psychoanalytic experience in a session.[39] While dreams, with their visual content and verbal analysis, are the "royal road" to the unconscious, "the nonverbal essence of music travels an aural road to the same destination ... [T]he qualities of *music itself* provide important *points of entry* into unconscious processes."[40] She suggests that the unconscious has an inherent quality that is exquisitely attuned to sounds that include both words and music; this in part explains how music has the capacity to create and recreate perspectives and affects formerly inaccessible. She herself, being musically trained, describes how one can think and feel in music, and that music can also function as a transitional object, providing comfort and also stimulation, as well as providing a unique aural entry into mental life.[41]

In one clinical example,[42] Nagel describes how Ms. A had struggled to find words to express feelings towards her and that had been evoked in sessions. Then in one session she associated to an early memory in which she tenderly recalled snow glistening under evening lights, which then led to associations about her love of sparkling jewellery and then to recall the melody from "Twinkle, Twinkle, Little Star". At that point her voice broke as she welled up in tears, as she recalled from early childhood a faint memory of her walking with her father on a snowy evening. Her father had left the family home soon after that time, and never reappeared. The patient began to hum the nursery rhyme, and then sing it, evoking the pain of the loss of her father and her yearnings for love—complicated by her mother's emotional unavailability. This proved an important mutative moment in the treatment.

Interestingly, Nagel herself also associated privately about an article she was then writing about Mozart around the time of his father's death in the summer of 1778. It was during that period that he composed his Piano Sonata in A minor, K. 310, the only piano sonata he wrote in that key and which conveys extreme emotional upheaval. A minor for Mozart was described by the musicologist Alfred Einstein as the "key of despair".[43] Mozart also composed at that time a set of variations on the children's song, "Ah, vous. Dirais-je, Maman", K. 265, to the tune of what we now sing as "Twinkle, Twinkle, Little Star". Hence the associative link between Ms. A's memory and that of the analyst.

Musical pieces do not just consist of abstract notes on the page, but those notes have to be interpreted. The same piece of music can be interpreted in various ways according to the emotional, physical, and technical tone that the interpreter gives them. The same could be said in some ways about the psychoanalyst's words. The same interpretation may be spoken in a variety of ways, depending upon the tone required at a particular moment of the analysis; as with Nagel's patient, they had reached a deep level of trauma that was evoked through the patient's surprisingly tender memories of her father.

Pinchas Noy[44] has suggested that there are three overlapping theories to explain the emotional effect of music. The first is the *narrative route*, when the audience identifies, through their imagination, the experience and the emotions appropriate to a narrative such as in opera, dance, and programme music. There are different kinds of narrative, the more obvious kind with an opera or oratorio, or in a religious service or communal activity; and specifically, musical narratives, such as the kind of drama of excitement, suspense, contradiction, and surprise seen in symphonic or chamber music, all based on purely instrumental means of expression. Creative musicians can tell a kind of story by manipulating and modifying the forms of music along the axis of time, or what I have called the musical journey.

Emotions are evoked by the *direct route* when the raw materials of music have a direct effect on the listener. This can occur first of all via representational means, much of which derives from the musical or non- or preverbal components of language. One can see this in, for example, the laughter of the orchestra accompanying the famous catalogue aria of Leporello in Mozart's *Don Giovanni*, or the weeping in Dido's lament from Purcell's *Dido and Aeneas*. For Noy, the specific power of artistic and musical communication lies in its ready access to preverbal memory and to its primary process organisation,[45] which links up with what I have already described both from the work of Max Graf, and also in the whole area of communicative musicality outlined in Chapter Two.

Noy suggests that music also has the power to arouse emotions in the listener in some direct ways that are not dependent upon any communication of meanings. He cites as an example the theory of isomorphism, as represented by Suzanne Langer and C. C. Pratt, who put forward the notion that the form of a musical phrase or motif is similar to that of the contours of an emotion, so that the structures of music and our emotions have the same essential shape. As a consequence of this similarity, they propose, we take music to be symbolic of our emotional life. Music then reflects the morphology of feeling, not representation as such, or anyway not a tight kind of representation. But this theory has a number of weaknesses: it is rather intellectual and cognitively based, and it does not allow for the fact that music can have a direct effect on our emotions.

Finally, for Noy the third route in which music can express emotions is that of the *indirect route via organisational activity*. This is when, as DeNora emphasises, that music provokes the listener to some kind of organisational activity where emotions are evoked and experienced, with greater or lesser amounts of psychic work, such as during participation in communal activities, whether that be in a concert or other social events.

I have already cited how Gilbert Rose, in his book *Between Couch and Piano* speculates from Freud's views on affect that the interplay of tension and release either generates or reflects affect, or is at least closely related to the nature of affect; and if a *dynamic of tension and release over time* lies at the heart of affect, a similar dynamic will lie at the heart of artistic endeavour, including that of music.[46] Thus, as Rose points out, Freud's theory of affects leads to the conclusion that music and affect share the *common dynamic of tension and release*.

One could then maintain that much of the emotional satisfaction in listening to tonal music comes from experiencing tonal tensions and their resolution without necessarily tying these movements to specific affects. Music then is a mirror of emotional tensions, as Suzanne Langer basically proposes in her theory of isomorphism.[47] However, this does not account for a vast amount of music where there are words which particularise the emotions, such as in choral music and lieder; or in many cultures where music and emotion are intimately linked, such as with the Kaluli people I described in Chapter One; or music which adds to or heightens specific emotions as in opera; or with a number of symphonic pieces, notably with Mahler, where the dramatic or narrative element also particularises the emotions. Movements of tension and release in music are a crucial element of musical expression, but cannot account substantially for the emotions both produced and evoked by music in listeners. One needs to take account of all these other features, and add to this the vital role of the artistic vision.

Musical listening

I began the book with some observations about the relationship between musical and psychoanalytic understanding and listening, as after all psychoanalysis is a listening discipline. I will end with further thoughts about the listening process. While of course there are significant differences between understanding, and listening to, a musical performance and a patient in a consulting room, I have maintained that there is also some common ground. In psychoanalytical listening, one is listening simultaneously to the "surface" and the "depth" of the patient's communications, to both the conscious and underlying unconscious stream of thoughts and feelings. Analytic listening, however intellectually taxing at times, also entails a responsive, receptive, or affective kind of listening, more like trying to make sense of the shape of the communications. During a session the analyst may become immersed in the flow of the patient's material, in a kind of musical "reverie",[48] which can arise in the analyst particularly during intense emotional exchanges. One has to listen to the patient's "music", their underlying communicative musicality, the flow of unconscious thoughts and feelings.

As I have suggested in the Overture, every patient has their own music, but every analyst and patient encounter creates a music of its own. The analyst is thus engaged with listening to both the patient and to themselves, and to their own responses to what the patient brings. One could say there are different kinds of music going on in the analytic situation. The patient may attempt to communicate an inarticulate fantasia, or maybe a suspiciously articulate composition. Then there is music that goes on in the analyst as they listen, and in the way that interpretations are composed. And then there is the music made jointly in the analysis, a form of duet.

There is an internal listening, involving a certain kind of receptivity to the unconscious, which seems to have parallels with listening to music. It is, as Parsons also points out,[49] that kind of receptiveness which comes in when listening to poetry. This is similar to what Reik described as "hearing oneself", by which he meant the analyst using their unconscious as a receiving apparatus.[50] Parsons describes how being receptive to the "internal" music aroused in a listening analyst helps the analyst understand the external music that is the patient.

In trying to explore this listening process, it may help to consider the complexities of musical listening where of course dissonance and consonance are in constant interaction. Listening to music, being affected by the music one hears, involves a number of different elements; there is a network of human subjects engaged in complex acts of communication and interpretation involving both the intellect and the emotions. Furthermore, as Jessica Holmes shows, experience from those who are deaf in different ways reveals that "listening encompasses a full spectrum of sensory experiences, musical contexts, individual preferences, cultural practices, and social experiences—which amount to an ever-evolving set of listening states".[51] In short, listening is a holistic experience.

Holmes cites the experience of the percussionist Evelyn Glennie as an "expert listener". Despite being unable to hear sound below 91 decibels, Glennie is a world-class performer by means of what she describes as a technique of "touching the sound". She engages her whole body as a resonating chamber, by which to sense and digest sounds, while simultaneously integrating visual cues, movement, and imagination.[52] It is not a question of a deaf person using other sensory modalities than hearing to play music as a way of compensating for loss, the traditional and prejudiced way of describing such talents, but that she brings into focus that sound itself is a multisensory experience. As Glennie puts it, "Listening is about more than just hearing; it is about engaging, empowering, inspiring and creating bonds. True listening is a holistic act."[53] Thus, an understanding of the complexities of deafness deepens one's understanding of music's social, relational, and material contours, and how musical listening makes use of multiple sensory modalities.

Musical listening is more than ordinary day-to-day listening, but it is not so clear about what kind of form this might take. Most people have a musical sense, apart from those with congenital amusia. Most people can enjoy listening to music, without specialised knowledge, in the same way that an appreciation of a piece of theatre or of a poem does not require specialist knowledge, though that knowledge may deepen the understanding and potentially the enjoyment of the work. But even the non-specialist will be listening to a piece of music in certain ways, however unconsciously. I would add that a capacity to listen is essential in understanding not only the meaning of music but the meaning of the other's communications.

Eric Clarke[54] describes how listening to music is an example of tuning in to our environment. His approach, as I mentioned in Chapter Four, influenced by the thought of James Gibson, is "ecological", that is, it takes as its central principle the relationship between a perceiver and their environment. Listening to music in this approach is the continuous awareness or perception of meaning. Rather than considering perception to be a constructive process in which the perceiver builds structure into an internal model of the world, the ecological approach emphasises the structure of the environment itself, and the active "resonance" or "tuning" of the perceptual system into constant, or "invariant", elements of the surrounding environment. This reflects in part the evolutionary origin of the perceptual apparatus as an adaptation to the subject's surroundings.

Listeners become particularly attuned to invariants, so that in musical listening a theme or a motif can be regarded as an invariant, a "pattern of temporal proportions and pitch intervals ... that is left intact, and hence retains its identity, under transformations such as pitch transposition or changes in global tempo".[55] Listeners can become attuned to certain invariants that specify a style or a pattern of consonance and dissonance, or a repeated pattern of musical elements. I also think that there is a basic tendency for listeners to look for a home base as an essential invariant. Performers have a particular approach to listening, captured in the notion of entrainment; the performer needs to be constantly listening to the other performers while adjusting and readjusting to the others' responses to them.

Clarke describes how the idea of invariance then leads on to the principle of "affordance", a term coined by Gibson to refer to how objects in the environment furnish the subject various properties for good or ill. Thus, food affords what is edible, other objects may afford danger. The child learns what objects afford, what to take, what to avoid. Affordance implies a complementarity between the subject and their environment. Music making in many cultures is an example of affordance between the subjects and their communities, though in the Western tradition, listening to music has become more autonomous.

When listening to music, I have already described how the subject is drawn into hearing movement in an imaginary or virtual space, or what for Clarke is a perceptual space. For Clarke, the sense of motion is fictional and a consequence of the fact that our auditory system is fine-tuned for detecting motion. When listeners perceive motion in music, they may experience a powerful identification with music if they find it pleasurable, or they may be turned off by the music if it's not their cup of tea. "The relationship between listener and music defines an aesthetic attitude, where the listener is both aware of what is going on in the music and what it might mean, and also has a sense of his or her own perspective on that meaning."[56] This entails what can be called a "subject-position" taken by the listener, and refers to the way in which the characteristics of the musical material shape the character of a listener's response or engagement, or non-engagement. One can say that music often solicits or even demands a particularly circumscribed response from the listener by means of its formal properties.

There are, as Clarke discusses, various ways of listening to music. Music in the Western art tradition has tended to become "autonomous", less connected to a clear social function as can be seen in, say the Kaluli people. Autonomous listening, then, is a kind of listening that follows the structure of an individual work but leaves out much of the music's social and cultural influences, though it has come to be a powerful way of listening. It involves a *virtual* environment into which the listener is drawn. Indeed, as Clarke shows, "A significant component of music's capacity to transform the listener lies in its power to temporally structure the sense of self. By virtue of its highly organized temporal but non-spatial structure, music provides a virtual environment in which to explore, and experiment with, a sense of identity."[57] This can be at the individual or group level; music can bring people together in powerful ways, enhancing group loyalties and identifications, or it can enable an individual to identify with a particular approach to life and art—be it an opera buff or a rap artist. One can see how music offers a virtual reality for such explorations, while also remaining firmly anchored in the everyday world.

The Italian psychoanalyst Antonio Di Benedetto has developed what one could call a creative theory of listening, using both musical and psychoanalytic influences. The first part of his book *Before Words* discusses the common ground between psychoanalytic and artistic thought, which he sees in the area of the unseen, the unheard and unsaid, located at the threshold of nascent thoughts. "Both the artist and the psychoanalyst aim at touching the germinative points of our mental processes, so that our mind can face, without being overwhelmed, those experiences, potentially disruptive, which are located at the border between integration and disintegration, chaos and feeling, the a-symbolic and the symbolic."[58]

Both the artist and the analyst need to listen to their unconscious in order to tap into inchoate ideas and feelings, which become the source of their formulations, as Max Graf showed when he described how the composer reaches down into their unconscious, tapping into childhood memories and experiences and external influences to produce their most important creative work. The artist is adept at incorporating unconscious influences into their creations.

In the second part of his book, Di Benedetto focuses on how to apply music to psychoanalysis, what music can teach analysts about how to listen to their patients creatively, to allow patient and analyst to reach inchoate areas of the mind and yet try to give them form, much as the artist, and in particular the composer does, the sound world being one which reaches the deep preverbal and non-verbal areas of the mind. Music can rend our psychoanalytic ears more sensitive, helping us to value more our sensitivity to the rhythms, harmonies, and the melodies of the mind, paying attention to subtle nuances of the voice, focusing on the messages unconsciously conveyed by the voice rather than only the content of speech.

Music, as he puts it, trains us to the

> unheard voices of the unconscious, those voices which, like a slip of the tongue, infiltrate language, bypassing its conscious elaboration. Moreover, music can stimulate the creation of alternative mental paths, suggest changes of route for our thoughts, and promote a *mobility* of ideas, which facilitates the crossing of less familiar psychic areas. Musical listening leads to lateral thinking and experiencing, so that we can capture peripherally things that we feel but are still unable to define. These are things about which we can say very little ... In short, music teaches us how *to listen to what we cannot say.*[59]

He adds that it is as though in the psychoanalytic encounter the analyst needs to be able to read the musical score in the unconscious of the patient and then translate it into listenable sound. This occurs through listening to the quality and inflections of the patient's voice as what is being said in the session; through the voice we are in touch with the patient on a basic bodily level, close to the unconscious. Because it is so important to listen to the music of the patient, whenever I conduct clinical seminars or offer clinical supervisions, I rarely read any written account of analytic sessions, as I find this distracts from hearing the musical quality of the patient's speech, which reaches down to the unconscious. Too much focus on the written account of a session can miss the patient's idiom.

Aaron Copland tackles how we listen to music by breaking the listening process into its component parts—the sensuous, expressive, and sheerly musical planes,[60] which can also be applied to analytic listening. Listening sensuously means listening for the sheer pleasure of the musical sound itself, when we hear music without thinking too much; what one could describe as "just listening". Of course, just listening may also be a painful affair, grating on the ears, and in an analytic session, that may be a useful communication about the patient's own conflicts, as I illustrated with the example from Francis Grier's patient in Chapter One.

For Copland, all music also has a certain meaning behind the notes, expressing what the music is all about; that meaning may shift with each new hearing. That is also certainly the case in analysis, where meanings shift and are overdetermined. Each new session, each new hearing of what may be similar issues, may bring a new dimension into consciousness.

The third and musical plane refers to the existence of the musical pitches themselves, and how they are manipulated by the composer and played by the performer; this is the technical aspect of music, about which listeners have varied amounts of knowledge. This is parallel to listening to the music of the patient, their tone, timbre, rhythms, and so on.

For Copland, the ideal listener creatively and actively combines all three forms of listening—they are both inside and outside the music simultaneously, judging it and enjoying it, "wishing it would go one way and watching it go another—almost like the composer at the moment he composes it; because in order to write his music, the composer must be inside and outside his music, carried away by it and yet coldly critical of it. A subjective and objective attitude is implied in both creating and listening to music."[61] This implies, one could add, the active participation of the listener, including the active involvement of their imagination responding to the power of the composer's musical vision.

Similarly, analytic listening implies not only being affected by the emotional atmosphere created in the session, but also examining objectively what is being said and reflecting on what is being evoked.

Marion Milner described a particular form of attentive listening in analysis which is in touch with the body at a deep level, and I think is related to musical listening; this form of listening is particularly necessary when aiming to treat borderline patients, who may have difficulties in symbolic functioning, but it can apply in other clinical situations. Her notion began when she used the term "concentration of the body" to describe certain situations to do with how one attends to an object being painted, a kind of "active stillness of waiting and watching that embraced both inner and outer, subject and object, sky and earth, in a unity which yet recognised duality".[62] For the making of her free drawings, she described a kind of spreading of the imaginative body in wide awareness, which included one's own body, as attention involves a sinking down into internal body awareness, not seeking for "correct" interpretations, and not even seeking for ideas, but allowing ideas to emerge spontaneously, if they are to do so at all.[63] Such wide listening to the body is different from free-floating attention, where one allows whatever comes to mind to appear, as this is at a different level, more at or towards the surface of the mind. The deeper listening is perhaps similar to what Evelyn Glennie describes as "touching the sound", when she engages her whole body as a resonating chamber.

This kind of wide or deep listening can also be described as "time-free hearing" in musical listening, when, as Anton Ehrenzweig describes, consciousness sinks to a lower level, not looking for whole forms, but allowing the mind to become unfocused and thereby becoming in touch with the "time free" unconscious.[64] As he points out, this is well expressed by Mozart who described the gradual disintegration of surface time in conceiving a new composition, which is then put down as a whole; he moves from time-free hearing to coherence. Mozart reports how first bits and crumbs of the piece join together in his mind, then the soul gets warmed to the work, and the thing grows more and more, and "I spread it out broader and clearer, and at last it almost gets finished in my head … All the inventing and making goes on in me as in a beautiful strong dream."[65]

Psychoanalytic listening then, like musical listening, is a complex and multi-sensorial activity, involving surface and deep layers of the mind, a special kind of "layered" listening. For the patient to free associate in the quiet space of the consulting room, they also need to develop what Ken Robinson has called the "silent inner chamber" in which to listen to the flow of their thoughts.[66] Robinson brought this notion to mind in considering John Cage's point that in an anechoic chamber, which is as silent as technologically possible, and without external distractions, the "mind is free to enter into the act of listening, hearing each sound just as it is, not as a phenomenon more or less approximating a preconception".[67] The consulting room, though not an anechoic chamber, is nonetheless a place generally free from the noise of everyday life, thereby facilitating a close internal listening, from the bodily to the intellectual level.

* * *

I have argued that psychoanalytical and musical listening share similar states of mind, experiences, and cognitive processes, particularly in the area of non-verbal communication. Music reaches into areas of experience where words may leave off; that makes it both deeply pleasurable, and in some ways deeply mysterious. Through the medium of enveloping sound, we are put in touch with feelings about our deep inner selves but also with beauty and the soul. The book has been an attempt to look into the nature of the musical experience, but the mystery remains.

Notes

Chapter One

1. Merriam, A. (1964, p. 31)
2. Cook, N. (1987)
3. Nass, M. (1971, p. 309)
4. Lombardi, R. (2008)
5. Reik, T. (1953, pp. 19–20)
6. Reik, T. (1953, p. 23)
7. Grier, F. (2019)
8. Nagel, J. (2013)
9. Nagel, J. (1993).
10. Parsons, M. (2014, p. 113)
11. Parsons, M. (2014, p. 158)
12. Reik, T. (1953, p. 12)
13. Blacking, J. (1976, p. 32)
14. Quoted in Brown, D. (2006, p. 204; italics in original)
15. Tchaikovsky, P. (1900, p. 15)
16. Sterba, R. (1965)
17. Benzon, W. (2001, p. 57)
18. Rouget, G. (1985)
19. Nettl, B. (2015, p. 260)
20. Kramer, L. (2011, p. 14)
21. Cooke, D. (1959, p. 16)

22. Schopenhauer, A. (1819, p. 264)
23. Magee, B. (2000, p. 171)
24. Hegel, G. (1886)
25. DeNora, T. (2000)
26. Zuckerkandl, V. (1973, pp. 2–3)
27. Gurney, E. (1880, pp. 380–381)
28. Eliot Gardiner, J. (2013)
29. *Sacred Music* (2008), Coro DVD, featuring The Sixteen
30. Feld, S. (1982)
31. Feld, S. (1982, p. 219)
32. Barenboim, D., & Said, E. (2002, p. 125)
33. Bostridge, I. (2015, p. 42)
34. Vogel, J. (1962, p. 6ff.)
35. Samson, J. (2016, p. 26)
36. Tanner, M. (1996)
37. Dahlhaus, C. (1971, p. 134)
38. Dahlhaus, C. (1971, p. 135)
39. Scruton, R. (2016, p. 34)
40. Solomon, R. (2007)
41. Cooke , D. (1959, p. 199)
42. Kramer, L. (2007, p. 29)
43. Starobinski, J. (2005, p. 77)
44. Dahlhaus, C. (1989)
45. Huron, D. (2006)
46. Segal, H. (1952, p. 203)
47. Cross, I. (2009)
48. Adorno, T. (1998, pp. 7–8)
49. Letter to Marshalk, quoted in Cooke, D. (1980, p. 54)
50. Blacking, J. (1976, pp. 51–52)
51. Tomlinson, G. (2015, p. 270)
52. Nussbaum, M. (2001)
53. Mahler, G. (1979, p. 346, letter to Bruno Walter)
54. Nussbaum, M. (2001, p. 269)
55. Jones, E. (1955, pp. 88–89)
56. Mitchell, D. (1975, p. 74)
57. Proust, M. (1923, p. 237)
58. Budd, M. (1985, p. 176)
59. Kennedy, R. (2014)
60. Bostridge, I. (2015, p. 21)
61. Barenboim, D., & Said, E. (2002, pp. 46–47)
62. Adès,T., & Service, T. (2012, p. 41)

63. Moore, J. N. (2004)
64. Sloboda, J. (1998)
65. Cross, I., & Morley, I. (2009)
66. Tippett, M. (1995, p. 15)
67. Leppert, R. (1993, p. 22)
68. Copland, A. (1985, p. 25)
69. Scruton, R. (1997, p. 353)
70. Benzon, W. (2001, p. 137)
71. Trevarthen, C. (1999)
72. Sievers, B., Polansky, L., Casey, M., & Wheatley, T. (2013)
73. Simpson, R. (1965, p. 9)
74. Simpson, R. (1965, p. 10)
75. Rosen, C. (1976, pp. 74–75)
76. Zuckerkandl, V. (1973, p. 144)
77. Scruton, R. (1997, p. 18)
78. Wigglesworth, M. (2018, p. 8)
79. Adès, T., & Service, T. (pp. 48–49)
80. Scruton, R. (1997, p. 46)
81. Sessions, R. (1950, p. 53)
82. Scruton, R. (1997, p. 63)
83. Moore, J. N. (1984, p. 62)
84. Moore, J. N. (2004, pp. 5–6)
85. Moore, J. N. (2004, p. 8)
86. Blacking, J. (1976, p. 24)
87. Blacking, J. (1976, p. 75)
88. Cone, E. (1974)
89. Bostridge, I. (2011, pp. 81–82)
90. Brendel, A. (2015, pp. 63–73)
91. McClary, S. (1991)
92. Boulez, P. (1963, p. 88)
93. Freud, S. (1915e)
94. Hall, M., & Rattle, S. (1996)
95. Adorno, T. (1971, p. 20)
96. Adorno, T. (1958, p. 39)
97. Reich, S. (2002, p. 36)
98. Rosen, C. (2010, p. ix)

Chapter Two

1. Anzieu, D. (1995, p. 174)
2. Anzieu, D. (1995, p. 186)

3. Rosolato, G. (1974)
4. Kramer, L. (2018, pp. 86–87)
5. Leppert, R. (1993, p. 29)
6. Kramer, L. (2018, p. 128)
7. Leppert, R. (1993, p. 29)
8. Green, A. (2003, p. 115)
9. Murray, L., Kempton, C., Woolgar, M., & Hooper, R. (1993)
10. Rosolato, G. (1969, p. 291)
11. Starobinski, J. (2005)
12. Dunn L., & Jones, N. (1994, p. 1)
13. Wigglesworth, M. (2018, p. 175)
14. Bertau, M.-C. (2007)
15. Bertau, M.-C. (2007, p. 143)
16. Kristeva, 2013, p. 307)
17. Laplanche, J. (1999)
18. Bakhtin, M. (1963, p. 6)
19. Kennedy, R. (2007, pp.74–75)
20. Bertau, M.-C. (2007, p. 153)
21. Gratier, M., & Trevarthen, C. (2007, p. 170)
22. Gratier, M., & Trevarthen, C. (2007, p. 170; italics in original)
23. Trainor, L., & Hannon, E. (2013)
24. Zentnor, M., & Eerola, T. (2010)
25. Trehub, S., & Nakata, T. (2001)
26. Malloch, S. (1999)
27. Trevarthen, C. (1999)
28. Gratier, M., & Trevarthen, C. (2007, p. 174)
29. Stern, D. (1985, p. 54ff.)
30. Langer, S. (1967)
31. Stern, D. (2010)
32. Stern, D. (2010, p. 4)
33. Stern, D. (2010, p. 42)
34. Proust, M. (1923, p. 237)
35. Stern, D. (2010, p. 21)
36. Stern, D. (2010, p. 53)
37. Stern, D. (2010, p. 84)
38. Kerman, J. (1985, p. 84)
39. Stern, D. (2010, p. 84)
40. Imberty, M. (2005, p. 337)
41. Schön, D. (1983, p. 54)
42. DeNora, T. (2000, pp. 78–79)
43. Kirschner, S., & Tomasello, M. (2010)

44. DeNora (2003, p. 1)
45. Cross, I. (2009, p. 189)
46. Cook, N. (2013, p. 411)
47. Bostridge, I. (2011, p. 111)
48. Walter, B. (1957, p. 112)
49. Service, T. (2012, p. 279)
50. Barenboim, D. (2008, p. 20)
51. Service, T. (2012, p. 22)
52. Service, T. (2012, p. 128)
53. Wagner, R. (1869)
54. Service, T. (2012, p. 165)
55. Wigglesworth, M. (2018, p. 16)
56. Wigglesworth, M. (2018, p. 44)
57. Wigglesworth, M. (2018, p. 47)
58. Wigglesworth, M. (2018, p. 49)
59. Small, C. (1998, p. 4)
60. Small, C. (1998, p. 9)
61. Small, C. (1998, p. 50)

Chapter Three

1. Peretz, I. (2006, p. 18)
2. Tomlinson, G. (2015, p. 252)
3. Lerdahl, F., & Jackendoff, R. (1983)
4. Scruton, R. (1997, p. 18)
5. Tomlinson, G. (2015, p. 12)
6. Pinker, S. (1997)
7. Brown, S. (2000, pp. 271–300)
8. Mithen, S. (2005)
9. Darwin, C. (1871, II, p. 337)
10. Miller, G. (2000, p. 349)
11. Miller, G. (2000, p. 342)
12. Brown, S. (2000, pp. 296–297)
13. Dissanayake, E. (2000, 2009)
14. Dissanayake, E. (2000, p. 404)
15. Smith, N., & Trainor, L. (2008)
16. Mithen, S. (2005)
17. Mithen, S. (2005, p. 192)
18. Wray, A. (2002)
19. Tomlinson, G. (2015)
20. Tomlinson, G. (2015, p. 65)

21. Tomlinson, G. (2015, p. 69)
22. Tomlinson, G. (2015, p. 48)
23. Tomlinson, G. (2015, p. 205; italics in original)
24. Tomlinson, G. (2915, p. 269)
25. Tomlinson, G. (2015, p. 288)
26. Cross, I., & Morley, I. (2009, p. 76)
27. Trehub, S. (2000)
28. Plantinga, J., & Trehub, S. (2013)
29. Trehub, S., Ghazban, N, & Corbeil, M. (2015)
30. Peretz, I. (2006, p. 18)
31. Peretz, I. (2013, pp. 551–564)
32. Peretz, I. (2006, p. 12)
33. Luria, A., Tsvetkova, L., & Futer, D. (1965)
34. Sacks, O. (2007, p. xiii)
35. Patel, A. (2008, p. 283ff.)
36. Peretz, I. (2006, p. 20)
37. Blacking, J. (1976, p. 9)
38. Nettl, B. (2015, p. 25)
39. Merriam, A. (1964, p. 27)
40. Trehub, S., Becker, J., & Morley, I. (2015)
41. Nettl, B. (2015, p. 35)
42. Balkwill, L.-L., Thompson, W. F., & Matsunaga, R. (2004); Fritz, T. et al. (2009)
43. Nettl, B. (2015, p. 35)
44. Trehub, S., Becker, J., & Morley, I. (2015)
45. Merriam, A. (1964, pp. 219–226)
46. Merriam, A. (1964, p. 227)
47. Freud, S. (1921c, p. 74)
48. Freud, S. (1921c, p. 116)
49. Rouget, G. (1985, p. 125ff.)
50. Durkheim, E. (1912)
51. Durkheim, E. (1912, pp. 217–218)
52. Rouget, G. (1985, p. xvii)
53. Rouget, G. (1985, p. 11)
54. Becker, J. (2004, p. 67)
55. Rouget, G. (1985, p. 73)
56. Rouget, G. (1985, p. 154)
57. Rouget, G. (1985, p. 326)
58. Becker, J. (2004)
59. Becker, J. (2004, p. 1)
60. Becker, J. (2004, p. 106)
61. Becker, J. (2004, p. 2)

62. Becker, J. (2004, p. 2)
63. Ehrenzweig, A. (1993, p. 35 ff.)
64. Bourdieu, P. (1990, p. 53)
65. Becker, J. (2004, p. 70)
66. Becker, J. (2004, p. 75)
67. Becker, J. (2004, p. 89)
68. Becker, J. (2004, p. 121ff.)
69. Feld, S. (1982, p. 3)
70. Feld, S. (1982, p. 20)
71. Feld, S. (1982, p. 33)
72. Feld, S. (1982, p. 45)
73. Feld, S. (1982, p. 161)
74. Feld, S. (1982, p. 219)

Chapter Four

1. Langer, S. (1955, p. 3)
2. Langer, S. (1955, p. 9)
3. Langer, S. (1942, p. 243; italics in original)
4. Freud, S., & Breuer, J. (1893a, p. 6)
5. Freud, S., & Breuer, J. (1893a, p. 6; italics in original)
6. Freud, S., & Breuer, J. (1893a, p. 8; italics in original)
7. Freud, S., & Breuer, J. (1893a, p. 17; italics in originala)
8. Freud, S. (1900, p. 460ff.)
9. Freud, S. (1900, p. 460)
10. Freud, S. 1918b, p. 23)
11. Freud, S. (1900a, pp. 461–462)
12. Freud, S. (1900a, p. 487; italics in original)
13. Freud, S. (1915d)
14. Freud, S. (1915e)
15. Freud, S. (1915d, p. 178; italics in original)
16. Freud, S. (1920g, p. 8; italics in original)
17. Freud, S. (1926d)
18. Freud, (1926d, p. 93)
19. Freud, S. (1940a, p. 146; italics in original)
20. Rose, G. (2004, p. 50)
21. Zuckerkandl, V. (1973, p. 144)
22. Langer, S. (1942)
23. Rayner, E. (1991, p. 34ff.)
24. Heimann, P. (1989, p. 74)
25. Heimann, P. (1989, p. 152)

26. McDougall, J. (1984)
27. McDougall, J. (1984, p. 388)
28. Marty, P., M'Uzan, M., & David, C. (1963)
29. Sandler, A., & Sandler, J. (1978, p. 292)
30. Sandler, A., & Sandler, J. (1978, p. 292; italics in original)
31. Klein, M. (1952, p. 81)
32. Klein, M. (1959, p. 251)
33. Green, A. (1999, p. 285; italics in original)
34. Green, A. (1999, p. 286)
35. Green, A. (1995 p. 211)
36. Green, A. (1999, p. 292)
37. Panksepp, J., & Trevarthen, C. (2009, p. 113)
38. Benvenuto, B., & Kennedy, R. (1986, pp. 167–168)
39. Habibi, A., & Damasio, A. (2014)
40. Damasio. A. (2010, pp. 23–24; italics in original)
41. Damasio A. (2010, p. 21; italics in original)
42. Damasio, A. (2018, pp. 107–108)
43. Damasio, A. (2018, p. 107)
44. Damasio, A. (2918, p. 108)
45. Damasio, A. (2010, pp. 24–25)
46. Damasio, A. (2018, p. 115)
47. Solms, M., & Nersessian, E. (1999); Solms, M. (2015)
48. Solms, M., & Nersessian, E. (1999, pp. 11–12)
49. Habibi, A., & Damasio, A. (2014, p. 92)
50. Panksepp, J., & Trevarthen, C. (2009, p. 120)
51. Koelsch, S. (2014)
52. Koelsch, S. (2014, pp. 170–171)
53. Sloboda. J. (1991)
54. Huron, D. (2006, p. 283)
55. Blood, A., & Zatorre, R. (2001)
56. Habibi, A., & Damasio, A. (2014, p. 96; italics in original)
57. Reik, T. (1936, p. 51ff.)
58. Reik, T. (1936, pp. 58–59)
59. Reik, T. (1936, p. 61)
60. Meyer, L. (1956)
61. Meyer, L (1956, pp. 14–15)
62. Meyer, L. (1956, p. 43)
63. Meyer, L. (1956, p. 155)
64. Meyer, L. (1956, p. 161)
65. Eliot Gardiner, J. (2013, p. 344)
66. Meyer, L. (1956, pp. 193–195)

67. Kennedy, R. (2007, p. 74)
68. Benzon, W. (2001)
69. Benzon, W. (2001, pp. 23–25)
70. Benzon, W. (2001, p. 57)
71. Benzon, W. (2001, p. 96)
72. Benzon, W. (2001, p. 196; italics in original)
73. Benzon, W. (2001, p. 126 ff.)
74. Benzon, W. (2001, p. 141)
75. Juslin, P. (2013)
76. Juslin, P. (2013, p. 240; italics in original)
77. Juslin, P. (2019)
78. Gibson, J. (1979)
79. Clarke, E. (2005, p. 203)
80. Clarke, E. (2005, p. 92)
81. DeNora, T. (2000 p. 5)
82. Willis, P. (1978)
83. DeNora (2000, p. 7)
84. DeNora (2000, p. 74)
85. DeNora (2000, p. 157)

Chapter Five

1. Juslin, P. (2019, p. 58)
2. Mitchell, D. (2007, p. 16)
3. Vygotsky, L. (1971, p. 211)
4. Levinson, J. (2015, p. 22ff.)
5. Levinson, J. (2015, p. 23)
6. Levinson, J. (2015, p. 26)
7. In Cochrane, T., Fantini, B., & Scherer, K. R. (2013, p. 57)
8. In Cochrane, T., Fantini, & Scherer (2013, p. 59)
9. Cooke, D. (1959, p. 31)
10. Kramer, L. (2011, p. 64)
11. Freud, S. (1919h, p. 225)
12. Christophers, H., & Mohr-Pietsch, S. (2019, p. 62)
13. Burnside, J. (2019, p. 10)
14. Kramer, L. (2007, p. 29)
15. D'Angour, A. (2015)
16. D'Angour, A. (2015, p. 189)
17. D'Angour, A. (2015, pp. 189–190)
18. McKeon, R. (1941, p. 1311)
19. Magee, B. (2000, pp. 83–101)

20. Magee, B. (2000, p. 87)
21. Magee, B. (2000, p. 91)
22. Magee, B. (2000, p. 90)
23. Schopenhauer, A. (1859, p. 448)
24. Schopenhauer, A. (1859, p. 448)
25. Schopenhauer, A. (1859, p. 456)
26. Schopenhauer, A. (1859, p. 455–456)
27. Magee, B. (2000, p. 208)
28. Higgins, K. (2011)
29. Higgins, K. (2001, p. xv)
30. Higgins, K. (2001, pp. xvi–xvii)
31. Higgins, K. (2001, p. 129)
32. Barenboim, D. (2008, p. 20)
33. Barenboim, D. (2008, p. 134)
34. Ridley, A. (2004, p. 1)
35. Kivy, P. (1990, p. 67)
36. Kivy, P. (1990, p. 202)
37. Hanslick, E. (1891, p. 28)
38. Hanslick, E. (1891, p. 49)
39. Ridley, A. (2004, p. 140)
40. Ridley, A. (2004, p. 160)
41. Ridley, A (2004, p. 162)
42. Kivy, P. (1990)
43. Cochrane, T., Fantini, B., & Scherer, K. R. (2013, p. 23)
44. Jankélévitch, V. (1983, pp. 58–59)
45. Jankélévitch, V. (1983, p. 1)
46. Jankélévitch, V. (1983, p. 74)
47. Jankélévitch, V. (1983, p. 57)
48. Scruton, R. (2009, p. 88)
49. Scruton, R. (1997, p. 93)
50. Scruton, R. (1997, p. 19)
51. Zuckerkandl, V. (1973, p. 87)
52. Adès, T., & Service, T. (2012, pp. 2–3)
53. Adès, T., & Service, T. (2012, p. 7)
54. Imberty, M. (2005, p. 63)
55. Service, T. (2012, p. 188)
56. Cooke, D. (1959, p. 17)
57. Cooke, D. (1959, pp. 75–76)
58. Cooke, D. (1959, p. 94)
59. Berlioz, H., & Strauss, R. (1948, p. 8)
60. Berlioz, H., & Strauss, R. (1948, p. 17)

61. Scruton, R. (2014, p. 144; italics in original)
62. Scruton, R. (2014, p. 146)
63. Scruton, R. (2018, p. 82)
64. Croce, B. (1913, p. 8)
65. Croce, B. (1913, p. 9)
66. Croce, B. (1913, p. 25)
67. Croce, B. (1913, pp. 33–34)
68. Scruton, R. (1997, p. 151; italics in original)
69. Scruton, R. (2018, p. 124; italics in original)
70. Kennedy, R. (2014, pp. 3–4)
71. Humphrey, N. (2011, p. 154)
72. Ward, K. (1998, p. 119)
73. Kennedy, R. (2014)
74. Freud, S. (1900a, p. 536)
75. Reik, (1953, p. 12)
76. Winnicott, D. W. (1971, p. 89; italics in original)
77. Winnicott, D. W. (1971, p. 5)
78. Winnicott, D. W. (1971, pp. 105–106)
79. Wigglesworth, M. (2018, p. 27)
80. Milner, M. (1987, p. 87ff.)
81. DeNora, T. (2003, p. 95)
82. DeNora, T. (2000, p. 47)
83. DeNora, T. (2003, pp. 99–104)
84. DeNora, T. (2003, p. 99)
85. Robinson, J. (2005, p. 293)
86. Robinson, J. (2005, p. 348)
87. Copland, A. (1958, p. 14)

Chapter Six

1. Nagel, J. (2018)
2. Feder, S., Karmel, R., & Pollock, G. (1990, 1993)
3. Di Benedetto, A. (2001); Rose, G. (2004); Imberty, M. (2005); Nagel, J. (2013)
4. Abrams, D. (1993)
5. Graf, M. (1947)
6. Graf, M. (1947, p. 23)
7. Graf, M. (1947, p. 25)
8. Graf, M. (1947, p. 5)
9. Graf, M. (1947 p. 80ff)
10. Graf, M. (1947, p. 98)
11. Graf, M. (1947, pp. 116–117)

12. Graf, M. (1947, pp.135–136)
13. Graf, M. (1947, p. 151)
14. Graf, M. (1947, p. 151)
15. Graf, M. (1947, p. 279; italics in original)
16. Graf, M. (1947, p. 377)
17. Graf, M. (1947, p. 453)
18. Roussillon, R. (2010)
19. Freud, S., & Breuer, J. (1895d, p. 288)
20. Freud, S. (1900a, pp. 277–278)
21. Freud, S. (1914g, p. 147)
22. Freud, S. (1914g, pp. 155–156)
23. Brenman Pick, I. (1985)
24. Roussillon, R. (2010, p. 1411)
25. Birksted-Breen, D., Flanders, S., & Gibeault, A. (2010, p. 40)
26. Leader, D. (2000, pp. 88–119)
27. Leader, D. (2000, p. 103)
28. Nass, M. (1993, pp. 21–40)
29. Nass, M. (1993, p. 28)
30. Sessions, R. (1950, p. 43)
31. Sessions, R. (1950, p. 44)
32. Sessions, R. (1950, p. 53)
33. Sessions, R. (1950, pp. 59–64)
34. Copland, A. (1939, p. 19)
35. Copland, A. (1939, p. 10)
36. Copland, A. (1939, pp. 21–23)
37. Copland, A. (1939, p. 25)
38. Cone, E. (1974)
39. Nagel, J. (2008)
40. Nagel, J. (2008, p. 513)
41. Nagel, J. (2013, pp. 3–4)
42. Nagel, J. (2008, p. 516)
43. Einstein, A. (1946, p. 244)
44. Noy, P. (1993)
45. Noy, P. (1993, p. 137)
46. Rose, G. (2004, p. 50)
47. Langer, S. (1942)
48. Lombardi, R. (2008)
49. Parsons, M. (2014, p. 158)
50. Reik, T. (1953, p. 12)
51. Holmes, J. (2017, p. 212)
52. Holmes, J. (2017, p. 171)

53. Holmes, J. (2017, p. 190)
54. Clarke, E. (2005)
55. Clarke, E. (2005, p. 35)
56. Clarke, E. (2005, p. 51)
57. Clarke, E. (2005, pp. 148–149)
58. Di Benedetto, A. (2001, p. 42)
59. Di Benedetto, A. (2001, p. 175; italics in original)
60. Copland, A. (1939, p. 7)
61. Copland, A. (1939, p. 14)
62. Milner, M. (1950, p. 93)
63. Milner, M. (1987, p. 240)
64. Ehrenzweig, A. (1953, p. 96ff.)
65. James, W. (1890, Vol 1., p. 255n.)
66. Robinson. K. (2015)
67. Cage, J. (1978, p. 23)

References

Abrams, D. (1993). Freud and Max Graf. In: S. Feder, R. Karmel, & G. Pollock (Eds.), *Psychoanalytic Explorations in Music* (pp. 279–307). Madison, CT: International Universities Press.

Adès, T., & Service, T. (2012). *Full of Noises*. London: Faber and Faber.

Adorno, T. (1958). *Philosophy of Modern Music*. A. Mitchell & W. Blomster (Trans.). New York: Continuum, 2004.

Adorno, T. (1971). *Mahler*. E. Jephcott (Trans.). Chicago, IL: University of Chicago Press, 1992.

Adorno, T. (1998). *Beethoven*. E. Jephcott (Trans.). Cambridge: Polity Press.

Anzieu, D. (1995). *The Skin-Ego*. N. Segal (Trans.). London: Karnac, 2016.

Bakhtin, M. (1963). *Problems of Dostoevesky's Poetics*. C. Emerson (Trans.). Minneapolis, MN: University of Minnesota Press.

Balkwill, L.-L., Thompson, W. F., & Matsunaga, R. (2004). Recognition of emotion in Japanese, Western and Hindustani music by Japanese listeners. *Japanese Psychological Research*, 46(4): 337–349.

Barenboim, D. (2008). *Everything Is Connected*. London: Weidenfeld & Nicolson.

Barenboim, D., & Said, E. (2002). *Parallels and Paradoxes*. London: Bloomsbury.

Becker, J. (2004). *Deep Listeners: Music, Emotion and Trancing*. Bloomington, IN: Indiana University Press.

Benvenuto, B., & Kennedy, R. (1986). *The Works of Jacques Lacan: an Introduction*. London: Free Association.

Benzon, W. (2001). *Beethoven's Anvil: Music in Mind and Culture*. Oxford: Oxford University Press.

Berlioz, H., & Strauss, R. (1948). *Treatise on Instrumentation*. T. Front (Trans.). New York: Dover, 1991.

Bertau, M.-C. (2007). On the notion of voice. *International Journal for Dialogical Science*, 2(1): 133–161.

Birksted-Breen, D., Flanders, S., & Gibeault, A. (2010). *Reading French Psychoanalysis*. New York: Routledge.

Blacking, J. (1976). *How Musical Is Man?* London: Faber and Faber.

Blood, A., & Zatorre, R. (2001). Intensely pleasurable responses to music correlate with activity in brain regions implicated in reward and emotion. *Proceedings of the National Academy of Sciences of the United States of America*, 98: 11818–11823.

Bostridge, I. (2011). *A Singer's Notebook*. London: Faber and Faber.

Bostridge, I. (2015). *Schubert's Winter Journey*. London: Faber and Faber.

Boulez, P. (1963). *Boulez on Music Today*. S. Bradshaw & R. R. Bennett (Trans.). London: Faber and Faber, 1971.

Bourdieu, P. (1990). Structures, habitus, practices. In: *The Logic of Practice*. Cambridge: Polity.

Brendel, A. (2015). *Music, Sense and Nonsense*. London: The Robson Press.

Brown, D. (2006). *Tchaikovsky, the Man and His Music*. London: Faber and Faber.

Brown, S. (2000). The "Musilanguage" model of music evolution. In: N. Wallin, B. Merker, & S. Brown (Eds.), *The Origins of Music* (pp. 271–300). Cambridge, MA: Massachusetts Institute of Technology Press.

Budd, M. (1985). *Music and the Emotions*. London: Routledge.

Burnside, J. (2019). *The Music of Time*. London: Profile.

Cage, J. (1978). *Silence, Lectures and Writings*. London: Marion Boyars.

Christophers, H., & Mohr-Pietsch, S. (2019). *A New Heaven*. London: Faber and Faber.

Clarke, E. (2005). *Ways of Listening*. Oxford: Oxford University Press.

Cochrane, T., Fantini, B., & Scherer, K. R. (Eds.) (2013). *The Emotional Power of Music*. Oxford: Oxford University Press.

Cone, E. (1974). *The Composer's Voice*. Berkeley, CA: University of California Press.

Cook, N. (1987). *A Guide to Musical Analysis*. Oxford: Oxford University Press.

Cook, N. (2013). *Beyond the Score*. Oxford: Oxford University Press.

Cooke, D. (1959). *The Language of Music*. Oxford: Oxford University Press.

Cooke, D. (1980). *Gustav Mahler: An Introduction to His Work*. London: Faber and Faber.

Copland, A. (1985) *What to Listen for in Music*. New York: Signet Classics.

Croce, B. (1913). *Guide to Aesthetics*. Indianapolis, IN: Bobbs-Merrill.

Cross, I. (2009). The evolutionary nature of musical meaning. *Musicae Scientiae*, 13(2): 179–200.

Cross, I., & Morley, I. (2009). The evolution of music. In: S. Malloch & C. Trevarthen (Eds.), *Communicative Musicality* (pp. 61–81). Oxford: Oxford University Press.

Dahlhaus, C. (1971). *Richard Wagner's Music Dramas*. M. Whittall (Trans.). Cambridge: Cambridge University Press, 1979.

Dahlhaus, C. (1989). *The Idea of Absolute Music*. R. Lustig (Trans.). Chicago, IL: Chicago University Press.

Damasio, A. (2010). *Self Comes to Mind*. London: William Heinemann.

Damasio, A. (2018). *The Strange Order of Things*. New York: Pantheon.

D'Angour, A. (2015). Sense and sensation in music. In: P. Destreé & P. Murray (Eds.), *A Companion to Ancient Aesthetics* (pp. 188–203). London: John Wiley.

Darwin, C. (1871). *The Descent of Man*. Princeton, NY: Princeton University Press, 1981.

DeNora, T. (2000). *Music in Everyday Life*. Cambridge: Cambridge University Press.

DeNora, T. (2003). *After Adorno, Rethinking Music Sociology*. Cambridge: Cambridge University Press.

Di Benedetto, A. (2001). *Before Words: Psychoanalytic Listening to the Unsaid through the Medium of Art*. G. Antinucci (Trans.). London: Free Association.

Dissanayake, E. (2000). Antecedents of the temporal arts in early mother–infant interaction. In: N. Wallin, B. Merker, & S. Brown (Eds.), *The Origins of Music* (pp. 389–410). Cambridge, MA: MIT Press.

Dunn, L., & Jones, N. (1994). *Embodied Voices*. Cambridge: Cambridge University Press.

Durkheim, E. (1912). *The Elementary Forms of Religious Life*. K. Fields (Trans.). New York: Free Press, 1995.

Ehrenzweig, A. (1953). *The Psycho-analysis of Artistic Vision and Hearing*. London: Routledge.

Ehrenzweig, A. (1993). *The Hidden Order of Art*. London: Weidenfeld & Nicolson.

Einstein, A. (1946). *Mozart*. London: Cassell.

Eliot Gardiner, J. (2013). *Music in the Castle of Heaven*. London: Allen Lane.

Feld, S. (1982). *Sound and Sentiment*. Durham, NC: Duke University Press.

Freud, S., with Breuer, J. (1893a). On the psychical mechanism of hysterical phenomena: preliminary communication. *S. E.*, *2*: 3–17. London: Hogarth.

Freud, S., with Breuer, J. (1895d). *Studies on Hysteria*. *S. E.*, *2*: 3–251. London: Hogarth.

Freud, S. (1900a). *The Interpretation of Dreams*. *S. E.*, *5*. London: Hogarth.

Freud, S. (1914g). Remembering, repeating and working–through (further recommendations on the technique of psycho-analysis, II. *S. E.*, *12*: 146–156. London: Hogarth.

Freud, S. (1915d). Repression. *S. E.*, *14*: 143–158. London: Hogarth.

Freud, S. (1915e). The unconscious. *S. E.*, *14*: 166–215. London: Hogarth.

Freud, S. (1918b). From the history of an infantile neurosis. *S. E.*, *17*: 3–122. London: Hogarth.

Freud, S. (1919h). The uncanny. *S. E.*, *17*: 219–256. London: Hogarth.

Freud, S. (1920g). *Beyond the Pleasure Principle. S. E.*, *18*: 3–64. London: Hogarth.

Freud, S. (1921c). *Group Psychology and the Analysis of the Ego. S. E.*, *18*: 74, 116. London: Hogarth.

Freud, S. (1926d). *Inhibitions, Symptoms and Anxiety. S. E.*, *20*: 77–175. London: Hogarth.

Freud, S. (1940a). *An Outline of Psycho-Analysis. S. E.*, *23*: 141–207. London: Hogarth.

Fritz, T., Jentschke, S., Gosselin, N., Sammler, D., Peretz, I., Turner, R., Friederici, A. D., & Koelsch, S. (2009). Universal recognition of three basic emotions in music. *Current Biology*, *19*(7): 573–576.

Gibson, J. (1979). *The Ecological Approach to Visual Perception*. Hillsdale, NJ: Lawrence Erlbaum.

Graf, M. (1947). *From Beethoven to Shostakovich: The Psychology of the Composing Process*. New York: Greenwood, 1969.

Gratier, M., & Trevarthen, C. (2007). Voice, vitality and meaning. *International Journal of Dialogical Science*, *2*(1): 169–181.

Green, A. (1995). Affects versus representations or affects as representations. *British Journal of Psychotherapy*, *12*(2): 208–211.

Green, A. (1999). *The Fabric of Affect*. London: Routledge.

Green, A. (2003). *Diachrony in Psychoanalysis*. London: Free Association.

Grier, F. (2019). Musicality in the consulting room. *Bulletin of the British Psychoanalytical Society*, *2*: 22–40.

Gurney, E. (1880). *The Power of Sound*. London: Smith, Elder.

Habibi, A., & Damasio, A. (2014). Music, feelings and the human brain. *Psychomusicology: Music, Mind and Brain*, *24*(1): 92–102.

Hall, M., & Rattle, S. (1996). *Leaving Home*. London: Faber and Faber.

Hanslick, E. (1891). *The Musically Beautiful*. G. Payzant (Trans.). Indianapolis, IN: Hackett.

Hegel. F. (1886). *Introductory Lectures on Aesthetics*. R. Bosanquet (Trans.). London: Penguin, 1993.

Heimann, P. (1989). *About Children and Children-no-longer*. London: Routledge.

Higgins, K. (2011). *The Music of Our Lives*. New York: Lexington.

Holmes, J. (2017). Expert listening beyond the limits of hearing: music and deafness. *Journal of the American Musicological Society*, *70*(1): 171–220.

Humphrey, N. (2011). *Soul Dust: The Magic of Consciousness*. London: Quercus.

Huron, D. (2006). *Sweet Anticipation*. Cambridge, MA: MIT Press.

Imberty, M. (2005). *La Musique Creuse Le Temps*. Paris: L'Harmattan.

James, W. (1890). *The Principles of Psychology*. New York: Dover, 1950.

Jankelevitch, V. (1983). *La Musique et L'eneffable*. C. Abbate (Trans.). Princeton, NJ: Princeton University Press, 2003.

Jones, E. (1955). *Sigmund Freud, Life and Work, Volume Two*. London: Hogarth.

Juslin, P. (2013). From everyday emotions to aesthetic emotions: towards a unified theory of musical emotions. *Physics of Life Reviews*, 10: 235–266.

Juslin, P. (2019). *Musical Emotions Explained*. Oxford: Oxford University Press.

Kennedy, R. (2007). *The Many Voices of Psychoanalysis*. London: Routledge.

Kennedy, R. (2014). *The Psychic Home*. London: Routledge.

Kerman, J. (1985). *Musicology*. London: Fontana.

Kirschner, S., & Tomasello, M. (2010). Joint music making promote prosocial behaviour in 4-year-old children. *Evolution and Human Behavior*, *31*(5): 354–364.

Kivy, P. (1990). *Music Alone*. Ithaca, NY: Cornell University Press.

Klein, M. (1952). Some theoretical conclusions regarding the emotional life of the infant. In: *Envy and Gratitude and Other Works* (pp. 61–93). London: Hogarth, 1980.

Klein, M. (1959). Our adult world and its roots in infancy. In: *Envy and Gratitude and Other Works* (pp. 247–263). London: Hogarth, 1980.

Koelsch, S. (2014). Brain correlates of music-evoked emotions. *Nature Reviews Neuroscience*, *15*: 170–180.

Kramer, L. (2007). *Why Classical Music Matters*. Berkeley, CA: California University Press.

Kramer, L. (2011). *Interpreting Music*. Berkeley, CA: California University Press.

Kramer, L. (2018). *The Hum of the World, a Philosophy of Listening*. Oakland, CA: California University Press.

Kristeva, J. (2013). *Passions of Our Time*. L. D. Kritzman (Ed.), C. Borde & S. Malovny-Chevallier (Trans.). New York: Columbia University Press, 2018.

Langer, S. (1942). *Philosophy in a New Key*. Cambridge, MA: Harvard University Press.

Langer, S. (1955). *Feeling and Form*. New York: Charles Scribner's Sons.

Langer, S. (1967). *Mind: an Essay in Human Feeling*. Baltimore, MD: Johns Hopkins University Press.

Laplanche, J. (1999). *Essays on Otherness*. London: Routledge.

Leader, D. (2000). *Freud's Footnotes*. London: Faber and Faber.

Leppert, R. (1993. *The Sight of Sound*. Berkeley, CA: University of California Press.

Lerdahl, F., & Jackendoff, R. (1983). *A Generative Theory of Tonal Music*. Boston, MA: MIT Press.

Levinson, J. (2015). *Musical Concerns*. Oxford: Oxford University Press.

Lombardi, R. (2008). Time, music and reverie. *Journal of the American Psychoanalytic Association*, *56*: 1191–1211.

Luria, A. R., Tsvetkova, L. S., & Futer, D. S. (1965). Aphasia in a composer. *Journal of Neurological Sciences*, *2*: 288–292.

Magee, B. (2000). *Wagner and Philosophy*. London: Allen Lane.

Mahler, G. (1979). *Selected Letters of Gustav Mahler*. A. Mahler & K. Martner (Eds.), E. Wilkins & E. Kaiser (Trans.). London: Faber and Faber.

Malloch, S. (1999). Mothers and infants and communicative musicality. In: I. Deliège (Ed.), *Rhythms, Musical Narrative* and *the Origins of Human Communication. Musicae Scientiae*, special issue, 1999–2000 (pp. 29–57).

Marty, P., M'Uzan, M. de, & David, C. (1963). *L'investigation Psychosomatique. Sept Observations Cliniques*. Paris: Presses Universitaires de France.

McClary, S. (1991). *Feminine Endings*. Minneapolis, MN: University of Minnesota Press.

McDougall, J. (1984). The "disaffected" patient: reflections on affect pathology. *Psychoanalytic Quarterly, 53*: 386–409.

McKeon, R. (Ed.) (1941). *The Basic Works of Aristotle*. New York: Random House.

Merriam, A. (1964). *The Anthropology of Music*. Evanston, IL: Northwestern University Press.

Meyer, L. (1956). *Emotion and Meaning in Music*. Chicago, IL: Chicago University Press.

Miller, G. (2000). Evolution of human music through sexual selection. In: N. Wallin, B. Merker, & S. Brown (Eds.), *The Origins of Music* (pp. 329–360). Cambridge, MA: MIT Press.

Milner, M. (1950). *On Not Being Able to Paint*. London: Heinemann.

Milner, M. (1987). *The Suppressed Madness of Sane Men*. London: Routledge.

Mitchell, D. (1975). *Gustav Mahler. The Wunderhorn Years*. London: Faber and Faber.

Mitchell, D. (2007). *Discovering Mahler, Writings on Mahler 1955–2005*. Woodbridge, UK: The Boydell Press.

Mithen, S. (2005). *The Singing Neanderthals*. London: Weidenfeld & Nicolson.

Moore, J. N. (1984). *Edward Elgar. A Creative Life*. Oxford: Oxford University Press.

Moore, J. N. (2004). *Elgar, Child of Dreams*. London: Faber and Faber.

Murray, L., Kempton, C., Woolgar, M., & Hooper, R. (1993). Depressed mothers' speech to their infants and its relation to infant gender and cognitive development. *Journal of Child Psychology and Psychiatry, 34*: 1083–1101.

Nagel, J. (1993). Stage fright in musicians: A psychodynamic perspective. *Bulletin of the Menninger Clinic, 57*(4): 492–503.

Nagel, J. (2008). Psychoanalytic perspectives on music: An intersection on the oral and aural road. *Psychoanalytic Quarterly, 77*(2): 507–530.

Nagel, J. (2013). *Melodies of the Mind*. London: Routledge.

Nagel, J. (2018). Memory slip: stage fright and performing musicians. *Journal of the American Psychoanalytic Association, 66*(4): 679–700.

Nass, M. (1971). Some considerations of a psychoanalytic interpretation of music. *Psychoanalytic Quarterly, 40*: 303–316.

Nass, M. (1993). The composer's experience: variations on several themes. In: S. Feder, R. Karmel, & G. Pollock (Eds.), *Psychoanalytic Explorations in Music* (pp. 21–40). Madison, CT: International Universities Press.

Nettl, B. (2015). *The Study of Ethnomusicology*, 3rd edition. Urbana, IL: University of Illinois Press.

Noy, P. (1993). How music conveys emotion. In: S. Feder, R. Karmel, & G. Pollock (Eds.), *Psychoanalytic Explorations in Music* (pp. 125-149). Madison, CT: International Universities Press.

Nussbaum, N. (2001). *Upheavals of Thought*. Cambridge: Cambridge University Press.

Panksepp, J., & Trevarthen, C. (2009). The neuroscience of emotion in music. In: S. Malloch & C. Trevarthen (Eds.), *Communicative Musicality* (pp. 105–146). Oxford: Oxford University Press.

Parsons, M. (2014). *Living Psychoanalysis*. London: Routledge.

Patel, A. (2008). *Music, Language, and the Brain*. Oxford: Oxford University Press.

Peretz, I. (2006). The nature of music from a biological perspective. *Cognition*, *100*(1): 1–32.

Peretz, I. (2013). The biological foundations of music: insights from congenital amusia. In: D. Deutsch (Ed.), *The Psychology of Music* (pp. 551–564). London: Elsevier.

Pick, I. B. (1985). Working through in the countertransference. *International Journal of Psychoanalysis*, 66: 157–166.

Pinker, S. (1997). *How the Mind Works*. New York: W. W. Norton.

Plantinga, J., & Trehub, S. (2013). Revisiting the innate preference for consonance. *Journal of Experimental Psychology: Human Perception and Performance*. doi: 10.1037/a0033471.

Proust, M. (1923). *The Prisoner*. C. Clark (Trans.). London: Allen Lane.

Rayner, E. (1991). *The Independent Mind in British Psychoanalysis*. London: Free Association.

Reich, S. (2002). *Writings on Music 1965–2000*. Oxford: Oxford University Press.

Reik, T. (1936). *Surprise and the Psycho-Analyst*. London: Routledge, 2014.

Reik, T. (1953). *The Haunting Melody*. New York: Farrar, Straus and Young.

Ridley, A. (2004). *The Philosophy of Music*. Edinburgh, UK: Edinburgh University Press.

Robinson, J. (2005). *Deeper than Reason*. Oxford: Oxford University Press.

Robinson, K. (2015). The ins and outs of listening as a psychoanalyst. *Empedocles: European Journal for the Philosophy of Communication*, 6: 169–174.

Rose, G. (2004). *Between Couch and Piano*. London: Routledge.

Rosen, C. (1976). *The Classical Style*. New York: W. W. Norton.

Rosen, C. (2010). *Music and Sentiment*. New Haven, CT: Yale University Press.

Rosolato, G. (1969). *Essais sur le symbolique*. Paris: Gallimard.

Rosolato, G. (1974). L'oscillation metaphor-metanomique. *Topique, 13*: 75–100.

Rouget, G. (1985). *Music and Trance*. Chicago, IL: Chicago University Press.

Roussillon, R. (2010). Working through and its various models. *International Journal of Psychoanalysis, 91*(6): 1405–1417.

Sacks, O. (2007). *Musicophilia*. New York: Random House.

Sacred Music (2008). Documentary series by BBC and The Open University featuring The Sixteen. Available on DVD from Coro.

Samson, J. (2016). The music of Oedipe. In Royal Opera House programme.

Sandler, J., & Sandler, A.-M. (1978). On the development of object relationships and affects. *International Journal of Psychoanalysis, 59*: 285–296.

Schön, D. (1983). *The Reflective Practitioner*. New York: Basic Books.

Schopenhauer, A. (1819). *The World as Will and Representation, Vol. 1*. E. Payne (Trans.). New York: Dover, 1969.

Schopenhauer, A. (1859). *The World as Will and Representation, Vol. 2*, 3rd edition. E. Payne (Trans.). New York: Dover, 1969.

Scruton, R. (1997). *The Aesthetics of Music*. Oxford: Oxford University Press.

Scruton, R. (2009). *Understanding Music*. London: Continuum.

Scruton, R. (2014). *The Soul of the World*. Princeton, NJ: Princeton University Press.

Scruton, R. (2016). *The Ring of Truth*. London: Allen Lane.

Scruton, R. (2018). *Music as an Art*. London: Bloomsbury Continuum.

Segal, H. (1952). A psychoanalytical approach to aesthetics. *International Journal of Psychoanalysis, 33*: 196–207.

Service, T. (2012). *Music and Alchemy*. London: Faber and Faber.

Sessions, R. (1950). *The Musical Experience of Composer, Performer, Listener*. New York: Atheneum.

Sievers, B., Polansky, L., Casey, M., & Wheatley, T. (2013). Music and movement share a dynamic structure that supports universal expression of emotions. *Proceedings of the National Academy of Sciences of the United States of America, 110*(1): 70–75.

Simpson, R. (1965). *Sibelius and Nielsen*. London: BBC Publications.

Sloboda, J. (1991). Music structure and emotional response: some empirical findings. *Psychology of Music, 19*: 110–120.

Sloboda, J. (1998). Music: where cognition and emotion meet. *The Psychologist, 12*: 450–455.

Small, C. (1998). *Musicking*. Middletown, CT: Wesleyan University Press.

Smith, N., & Trainor, L. (2008). Infant-directed speech is modulated by infant feedback. *Infancy, 13*(4): 410–420.

Solms, M. (2015). *The Feeling Brain*. London: Routledge.

Solms, M., & Nersessian, E. (1999). Freud's theory of affect: Questions for neuroscience. *Neuropsychoanalysis, 1*(1): 5–14.

Solomon, R. (2007). *True to Our Feelings*. Oxford: Oxford University Press.

Starobinski, J. (2005). *Enchantment, the Seductress in Opera*. C. J. Delogen (Trans.). New York: Columbia University Press, 2008.

Sterba, R. F. (1965). Psychoanalysis and music. *American Imago, 22*: 96–111.

Stern, D. (1985). *The Interpersonal World of the Infant*. New York: Basic Books.

Stern, D. (2010). *Forms of Vitality*. Oxford: Oxford University Press.

Tanner, M. (1996). *Wagner*. London: Harper Collins.

Tchaikovsky, P. (1900). *Guide to the Practical Study of Harmony*. Mineola, NY: Dover, 2005.

Tippett, M. (1995). *Tippett on Music*. M. Bowen (Ed.). Oxford: Oxford University Press.

Tomlinson, G. (2015). *A Million Years of Music*. New York: Zone.

Trainor, L., & Hannon, E. (2013). Musical development. In: D. Deutsch (Ed.), *The Psychology of Music*, 3rd edition. London: Elsevier.

Trehub, S. (2000). Human processing predispositions and musical universals. In: N. Wallin, B. Merker, & S. Brown (Eds.), *The Origins of Music* (pp. 427–448). Cambridge, MA: MIT Press.

Trehub, S., Becker, J., & Morley, I. (2015). Cross-cultural perspectives on music and musicality. *Philosophical Transactions of the Royal Society B, 370*: 20140096.

Trehub, S., & Nakata, T. (2001). Emotion and music in infancy. *Musicae Scientiae, 5*(1): 37–61.

Trevarthen, C. (1999). Musicality and the intrinsic motive pulse. *Musicae Scientiae*, special issue, 1999–2000 (pp. 155–215).

Vogel, J. (1962). *Leoš Janacek, His Life and Works*. London: Paul Hamlyn.

Vygotsky, L. (1971). *The Psychology of Art*. Cambridge, MA: MIT Press.

Wagner, R. (1869). *On Conducting*. E. Dannreuther (Trans.). New York: Dover, 1989.

Walter, B. (1957). *Of Music and Music-making*. London: Faber and Faber.

Ward, K. (1998). *In Defence of the Soul*. Oxford: Oneworld.

Wigglesworth, M. (2018). *The Silent Musician*. London: Faber and Faber. Willis, P. (1978). *Profane Culture*. London: Routledge.

Winnicott, D. W. (1971). *Playing and Reality*. London: Tavistock.

Wray, A. (2002). *The Transition to Language*. Oxford: Oxford University Press.

Zentner, M., & Eerola, T. (2010). Rhythmic engagement with music in infancy. *Proceedings of the American Academy of Sciences, 107*(13): 5768–5773.

Zuckerkandl, V. (1973). *Man the Musician*. Princeton, NJ: Princeton University Press.

Index